REDISCOVERING
ASTRONOMY

REDISCOVERING ASTRONOMY

by Eugene F. Provenzo, Jr.
and Asterie Baker Provenzo

designs and illustrations by
Peter A. Zorn, Jr.

OAK TREE PUBLICATIONS, INC.
PUBLISHERS, LA JOLLA, CALIFORNIA

First Edition
Manufactured in the United States of America
For information write to: Oak Tree Publications, Inc.
P.O. Box 1012, La Jolla, CA 92038

Library of Congress Cataloging in Publication Data

Provenzo, Eugene F
Rediscovering astronomy.

(Young inventor series; 1)
SUMMARY: Discusses the history of telescopes, and the scientists who played important parts in their development. Includes experiments.
1. Telescope—Juvenile literature. [1. Telescope] I. Provenzo, Asterie Baker, joint author. II. Zorn, Peter A. III. Title. IV. Series.
QB88.P76 522.2 80-17210
ISBN 0-916392-61-9

All experiments in this book are completely safe when properly conducted. However, some may require more caution than others. The publisher and authors have carefully detailed the necessary precautions that should be taken, and encourage the readers to fully understand all experiments prior to conducting them in order to better understand the specific phenomenon and to avoid any unnecessary mishap.

1 2 3 4 5 6 7 8 9 84 83 82 81 80

FOR ANN T. FREEDMAN AND
DAVID F. BAKER

Table of Contents to Experiments

ACKNOWLEDGMENTS

Our thanks go to the many people who have helped make REDISCOVERING ASTRONOMY possible. Helen Beibel, Mildred Merrick and Patricia Pardo of Richter Library, University of Miami, were particularly helpful in assisting our research. Helen Purdy of the Department of Archives and Special Collections, Richter Library, and M. Rosenfeld of the Dibner Library National Museum of History and Technology, took special care in helping us collect some of the illustrations. Dr. Joseph Barton, Dr. Robert Hendricks, Dr. Arnold Cheyney, Dr. Scott Baldwin and Dr. Lennie Middleton, School of Education and Allied Professions, University of Miami; Roy W. Carpenter, Communication Services, University of Miami; and Professor Philip Carter and Dr. Douglas Duke, Department of Physics and Astronomy, University of Miami, were all particularly helpful with technical assistance and suggestions for improving the book. John Power of Stern Brothers Plumbing Supply Company and Maury Gilbert of Insight Optical Manufacturing, Miami, were invaluable in providing technical information relating to the construction of the telescopes. Ann and Steven Freedman of New Haven, along with Liz Lasensky of Miami, provided their usual generous assistance. We would also like to thank the staff members at Oak Tree Publications, Inc., who have worked on this book for their sensitivity and professionalism in the production of this book.

REDISCOVERING ASTRONOMY is a book about the inventions and discoveries that have made **astronomy** what it is today. It is also a book about the people who have played an important part in the history of astronomy.

Throughout this book you will be able to experiment and re-create many of these discoveries and inventions just as they were done in the past. You will be able to "rediscover astronomy" by making your own rotating quadrant, sextant, Galilean refracting telescope, and a Newtonian reflecting telescope.

As the Swiss psychologist Jean Piaget has explained, "The essential functions of intelligence consist in understanding and inventing, in other words in building up structures by structuring reality." By rediscovering astronomy, you will come to understand how the history of astronomy is a reflection of the growth of our concept of the Earth, its place in the universe and of the universe itself.

Many of the astronomical instruments in this book can be made with things found around most homes. Some of them will require materials you have to buy at a hardware or plumbing store. It will also be necessary to obtain special lenses for some of the instruments you will be constructing. Complete, easy-to-follow instructions and diagrams are given for all of the experiments you will be re-creating and for the instruments you will be making. Patterns are also included for many of the instruments.

Once you have finished reading **REDISCOVERING ASTRONOMY**, you will have come to understand something about the people and the great discoveries that have made astronomy the modern science it is today. You will also have discovered how astronomy has given us new ways of understanding our own planet and the universe in which we live.

ASTRONOMY

The scientific study of the sun, moon, planets, stars and other heavenly bodies beyond the Earth. It involves the observation, calculation and theoretical interpretation of their compositions, positions, dimensions, distances, locations and evolutions. The word "astronomy" can be traced back to two Greek words, *astron,* meaning "a star," and *nomos,* meaning "distribution."

Symbolic illustration of a medieval astronomer trying to discover the secrets of the universe through the stars.

The **stars**, **comets**, **moon**, **planets** and even an occasional **eclipse** have captured people's curiosity and imagination for thousands of years. The heavens shined far brighter for ancient people than they do today for modern man. There were no street-lights to obscure the majesty of the night sky. People looked to the skies to find out the time of day, the direction they were traveling and when to plant their crops. They saw in the constellations many of their gods and goddesses and wondered what an afterlife would be like in the heavens.

Ancient people observed the stars without the aid of optical instruments such as the **telescope**. Everything they observed was with their naked eye. They saw a great deal—the moon, five planets, the sun and about 2,000 stars. Through careful observations, they discovered that the stars moved with a regularity that seemed to suggest an almost divine purpose or plan to the universe.

The study of the heavens, or cosmology, dominated almost every ancient culture. Chinese, Indians, Mesopotamians, Egyptians, early Native Americans and early Mayans all attempted to explain the universe in which they lived. Wherever early people were involved in agriculture they were also very concerned with a knowledge of the seasons and being able to compute time. Their early understanding of astronomy was usually developed as part of their religion.

Over 4,000 years ago, the ancient Egyptians divided the heavens into constellations, or groups of stars with recognizable shapes. According to them, the heavens were formed by the goddess Nut, who is portrayed bent over on all fours. She was held in place by the air god, Shu. Over Nut's body sailed the sun and the moon.

The Babylonians, whose empire in southwestern Asia was very powerful from about 2000 to 1000 B.C., pictured the heavens as being a solid dome supported by a mountain peak that was surrounded by water.

STARS

Any heavenly body that is not a comet, meteor or nebula and that shines by its own light. A star is really a hot body of gas that can be smaller than the Earth or up to several million times as large as the sun.

COMETS

A bright heavenly body that can be observed from Earth when its orbit passes close to the sun. Comets have a starlike body that can be as large as Earth. They also have a head which surrounds the body or nucleus and sometimes a tail that can be up to 100 million miles long. Comets travel around the sun in elliptical orbits. Each complete orbit is called a period.

MOON

A natural satellite of the Earth that rotates around it once every 29.5 days. It is kept in its elliptical orbit by the pull of the Earth's gravity. The moon is about 1/18 of the Earth's mass and is 2,160 miles in diameter.

PLANETS

Heavenly bodies that are not a comet or a meteor and that rotate around the sun or another star. Mercury, Venus, Mars, Jupiter, Saturn, Uranus, Neptune, Pluto and Earth are the main planets.

ECLIPSE

A complete or partial blocking of light passing from one heavenly body to another.

TELESCOPE

An instrument that makes distant objects appear nearer and larger. The word comes from the Greek words, *tele,* which means "far," and *skopein,* meaning "look at."

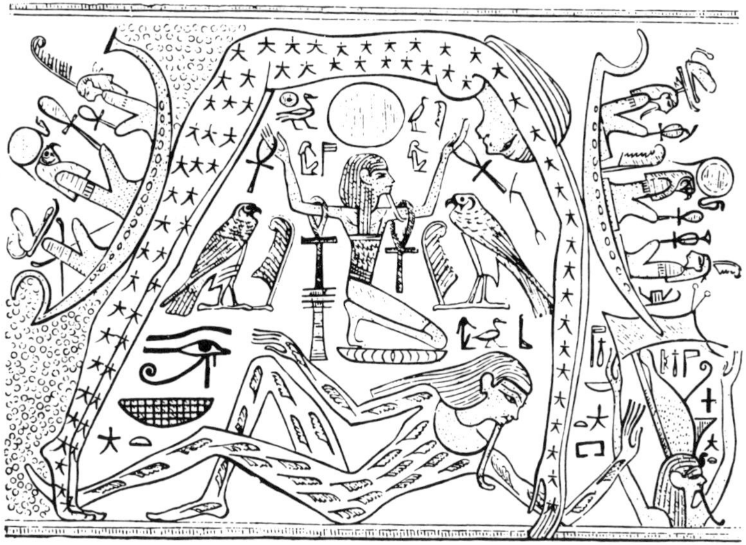

The goddess Nut creating the heavens.

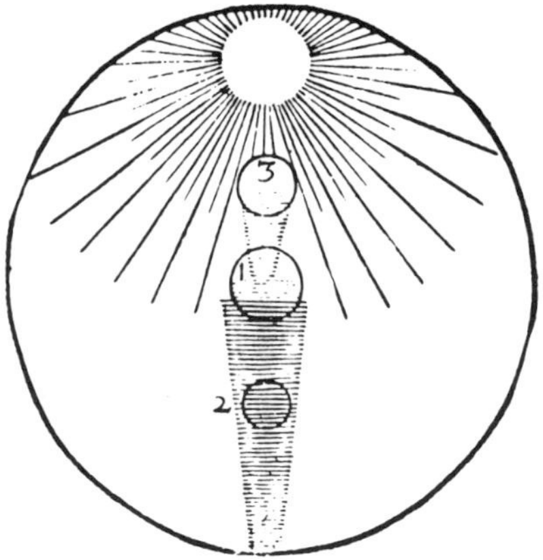

An illustration from Comenius' *Orbis Sensualium Pictus,* published in 1659, of a solar eclipse, in which the moon (3) is blocking the sun from the Earth (1) and a lunar eclipse, in which the moon (2) is in the Earth's shadow.

Records of the ancient Sumerians, dating from about 1500 B.C., describe a universe made up of eight spheres, each inside the other and each ruled by a different god.

All ancient people were concerned with being able to predict natural wonders like eclipses, which were considered very important omens. Almost all of these people had very primitive instruments to aid them, but some, like the Chinese, Egyptians and Babylonians, even had observatories for studying the stars. As soon as systems of numerals and writing were developed, ancient people began to make lists and compile elaborate tables of their observations of the heavens. At the point when they realized that certain events, like lunar eclipses, repeated themselves they began to study astronomy scientifically. Probably one of the most remarkable ancient observatories, where the movements of heavenly bodies were computed, was Stonehenge.

Mystery and magic have always surrounded Stonehenge. Stories dating back to the time of the legendary King Arthur (about A.D. 500) have described it as a city of the dead, or even a **Druid** altar for sacrifices. This massive monument is so old that its true history has long been lost. For hundreds of years, though, it has been suspected that Stonehenge was indeed an astronomical observatory and computer.

DRUID

An order of priests in ancient Gaul and Britain who appear in Irish and Welsh legends as sorcerers and prophets.

An illustration, from the late 1800s, of the ruins of Stonehenge. It shows a lone man leaning against the Heelstone.

Stonehenge is located on the Salisbury Plain in England. It is a 4,000-year-old observatory and temple constructed out of huge stones. The stones were arranged in such a way that Stonehenge probably actually worked as a computer to predict a **lunar eclipse**. It could also be used as a giant **astrolabe** to measure the altitude of the sun and stars.

LUNAR ECLIPSE

A lunar eclipse occurs when the moon enters the Earth's shadow or umbra.

ASTROLABE

During the Middle Ages the astrolabe was used as an instrument to find the altitude or angle of the sun, moon, planets and stars above the horizon. Astrolabes can also be used to determine the height of mountains, tall buildings or even the depth of wells. Before clocks were available, astrolabes were used to tell the time of day.

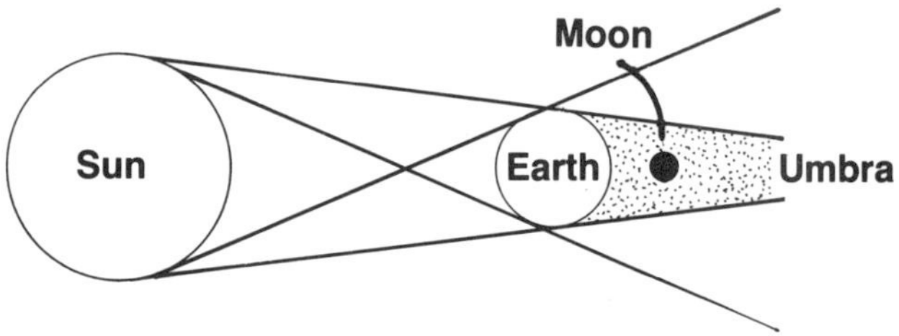

Lunar Eclipse

Aerial view of Stonehenge. The white spots mark the positions of the Aubrey Holes. The single standing stone next to the road is the Heelstone. *Dr. Harold Edgarton,* M.I.T., *Cambridge, Massachusetts*

Stonehenge was probably first laid out by **Neolithic** agricultural people around 2500 B.C. These people dug a circular ditch and then built a bank, 20 feet wide and as tall as a man, in the center. The bank was dug out on the northeast side to allow an unobscured view of the midsummer sunrise. A wooden post may have been used to mark this spot. Today, a huge sandstone boulder, known as the Heelstone, marks the line from the center toward the northeast direction of the midsummer sunrise.

NEOLITHIC

The latest cultural period of the Stone Age, which began about 10,000 B.C. with the invention of farming and the use of polished stone tools.

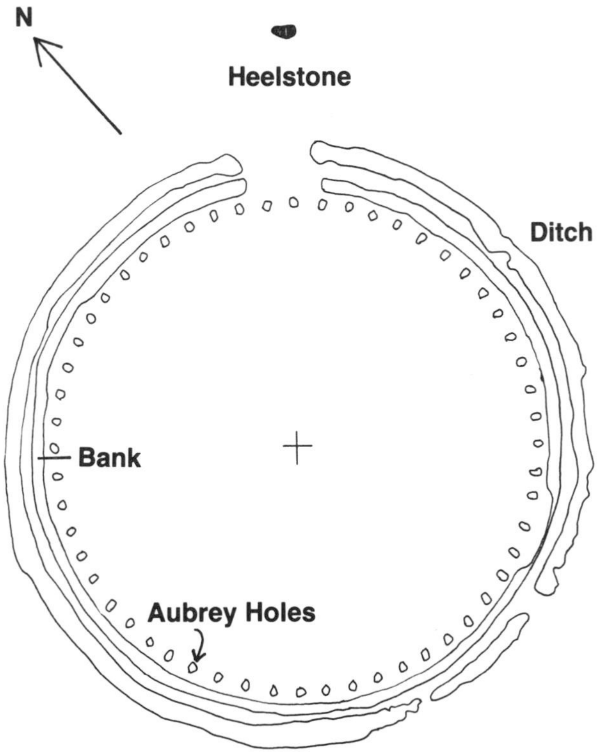

Diagram of Stonehenge showing the ditches, banks, Aubrey Holes and Heelstone of the earliest Stonehenge construction.

On top of the bank, 56 evenly spaced holes were dug in a circle. Known today as **Aubrey Holes**, they were originally used to keep track of the movement of the sun and the moon.

Evidently, in order to predict lunar eclipses, a system that involved the placement of stones was used at Stonehenge. The stones, representing the moon, the sun and the **nodes** of the moon, were shifted from one Aubrey hole to the next, according to the movement of the heavenly bodies they represented. In this way, they would have helped the people at Stonehenge calculate and predict lunar eclipses. Stonehenge could have also been used as a computer to keep track of the seasons, the time of the moonrise, patterns of the ocean's tides and even eclipses of the sun.

AUBREY HOLES

The 56 holes are named after John Aubrey, an Englishman, who wrote down his personal observations of Stonehenge during the 1800s.

NODES

The points where the sun and moon's orbits cross one another.

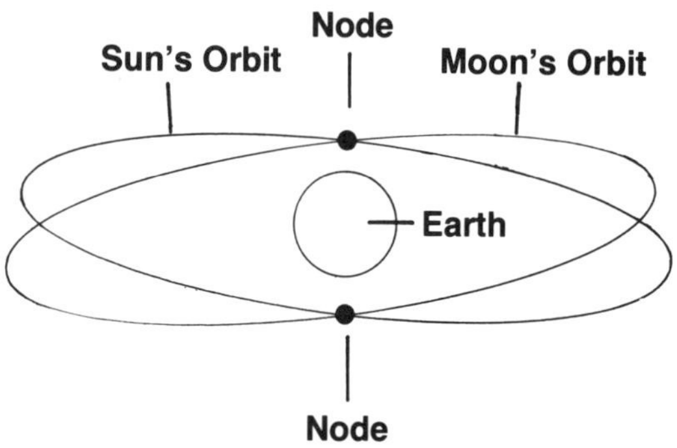

The Midsummer sun rising as seen from within the structure of Stonehenge. *From the CBS News Broadcast "Mystery of Stonehenge."*

If the ancient people who built Stonehenge did in fact use it as a computer, by moving the stones representing the moon, the sun and the nodes of the moon, then they had to know that the earth rotates, in a counter-clockwise direction, one complete orbit each day. The moon also moves in a counter-clockwise direction, but takes about 29.5 days to complete an orbit. So the sun and moon's orbits cross at the nodes of the moon, three times in 56 years. When their paths cross in this way, there is an eclipse of the moon which the Stonehenge computer was able to predict.

If the first Neolithic builders of Stonehenge worshipped the sun and moon as gods, then eclipses of the moon would have been great events in their lives. Anyone who could predict these eclipses would have been very powerful. Since total eclipses of the sun happen so rarely in the same place (about once every 300 years), it is unlikely that Neolithic people were very much concerned with predicting them.

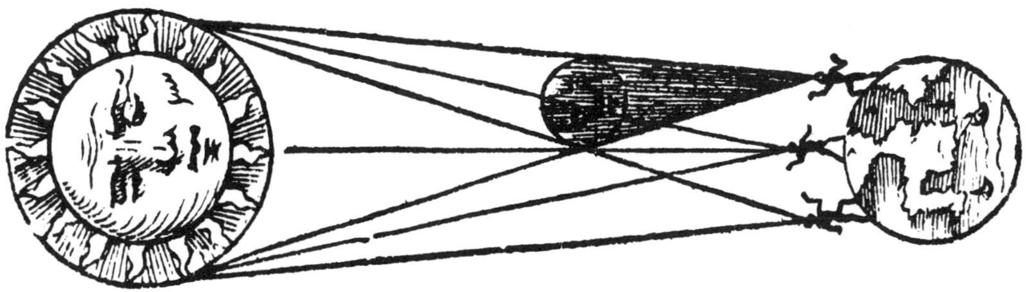

Diagram showing how three different people at different points on the Earth can see a total eclipse of the sun, a partial eclipse and no eclipse at all. From M. Blundevill's *A Plaine Treatise of the First Principles of Cosmographie,* published in 1594.

Stonehenge was rebuilt a number of times. The first Stonehenge was probably a simple arrangement of markers and wooden posts that was used before a system of writing numerals was invented to record astronomical observations and movements. By the time the huge stones were erected for the more recent Stonehenge III, about 500 years later, the site was probably used more for rituals and observations than as an actual computing machine.

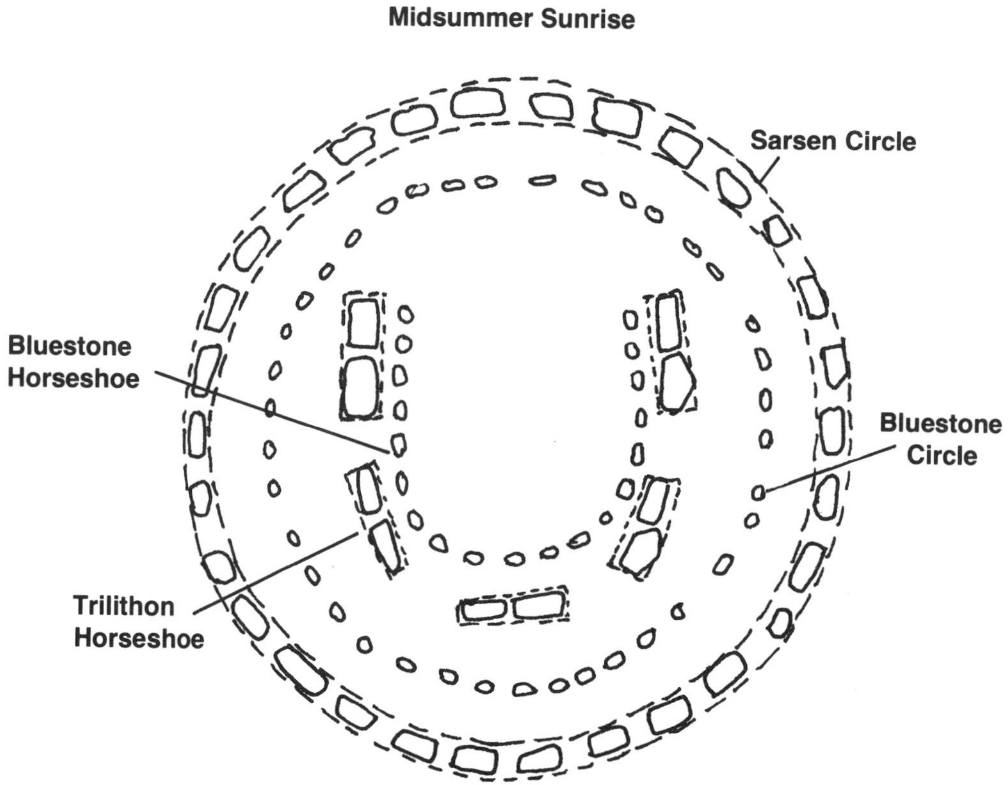

A plan of Stonehenge III showing the Sarsen Circle, the bluestone circle, the trilithons arranged in a horseshoe, and the bluestone horseshoe.

Stonehenge III was made up of a circle of 30 huge **sarsen** stones which stood upright inside the ring of Aubrey Holes. The stones in the Sarsen Circle were topped with lintels, or horizontal beams. Inside the Sarsen Circle stood five trilithons, also made of sarsen stone, arranged in a horseshoe formation. Within the trilithon horseshoe was another formation of smaller stones, known as bluestones. There was also a circle of bluestones between the Sarsen Circle and the trilithon horseshoe.

SARSEN

A hard, durable type of sandstone used for the stones and lintels at Stonehenge.

For centuries people have wondered how Stonehenge was built. Neither sarsen stones nor bluestones are found near the Salisbury Plain. The bluestones, which weigh up to five tons each, probably came from Wales, over 200 miles away. They may have been tied to wooden sledges and rolled along on log rollers. It would have taken one whole day for 20 men to move one stone a mile! The sarsen stones, which weigh up to 45 tons each, would have had to be moved over 20 miles to reach Stonehenge.

The huge stone circles of England's Stonehenge, and other very early monuments in Egypt, Babylonia and Mexico, tell us that ancient man was very concerned with observing the heavens and calculating time. In areas of the Mediterranean, where the sun shines most of the time, they used smaller instruments to help them plot and measure the movement of the sun. The most important of these instruments was the **sundial**.

SUNDIAL

An instrument used to measure local solar time. It measures the hour angle of the sun with a style, which casts the sun's shadow onto a calibrated dial.

The oldest known sundial was made in Egypt about 1500 B.C. It was constructed from stone in the shape of a flat bar about 12 inches long. A vertical T-shaped piece was attached to one end. The hours were distinguished by marks cut into the stone at uneven intervals. A plumb, or a small weight hung at the end of a line, was fitted to the sundial in order to make sure that it was level. The T-piece was turned toward the east in the morning and the west in the afternoon. The position of the shadow of the upper edge of the *T* determined the time of day.

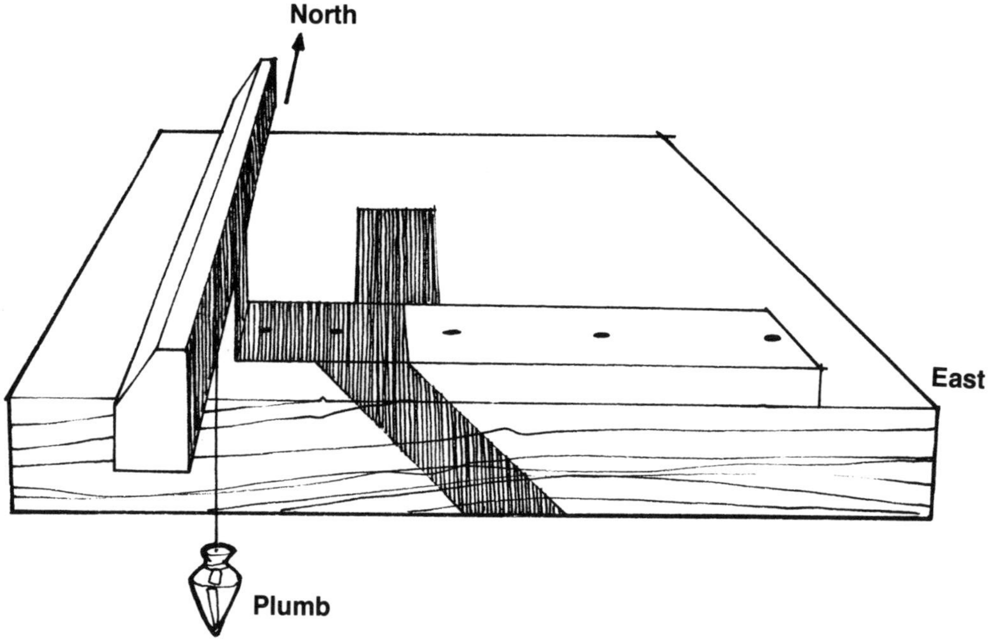

The oldest known sundial made in Egypt about 1500 B.C.

An illustration of sundials and other timekeeping devices from Comenius' *Orbis Sensualium Pictus,* published in 1659. Depicted are: (1) sundial, (2) shadow of the gnomon, (3) compass sundial, (4) hourglass, and (5) a clock.

It was not until 300 B.C. that any real improvements were made in the instruments used for observation in astronomy in the West. At that time the Greeks developed a more refined type of sundial known as the *gnomon.* The gnomon (pronounced ǹ ō män, which means "one who knows" in Greek) was simply a pointed vertical column erected on a horizontal plane. Besides being able to tell the time of day, the gnomon could also determine where north and south were. It is believed that the Chinese were using simple gnomons to measure the length of the sun's shadow and to determine the solstices as early as 3,500 years ago.

You can see how a gnomon works by pushing a straight stick, 2 to 3 feet long, into the ground at a right angle (90 degrees). Observe where the shadow is being cast by the stick. Wait until the beginning of an hour and set a small marker with the hour written on it at the top of the stick's shadow. Try to place markers for each hour of sunlight.

If you come back to your gnomon the next day and the sun is shining, you should be able to observe the shadow of the stick returning to the same points at the same time of day as when you positioned your markers. You have made a very simple but effective gnomon, or sundial.

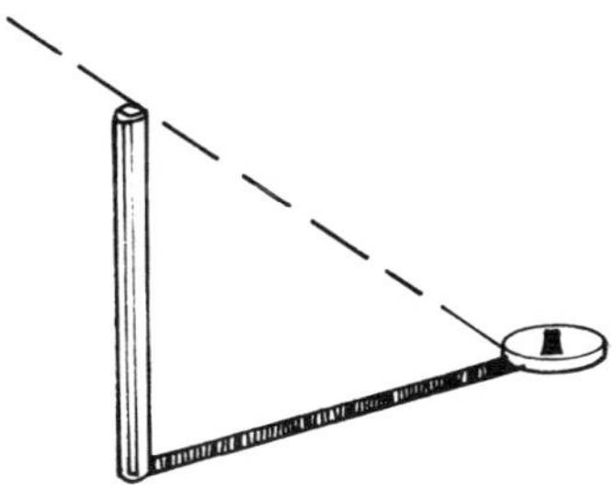

One o'clock

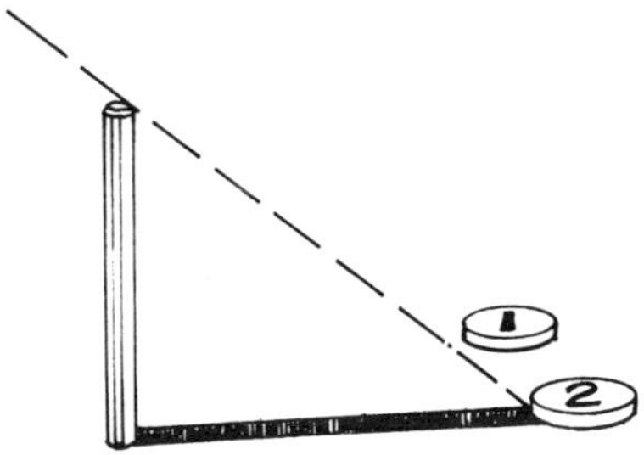

Two o'clock

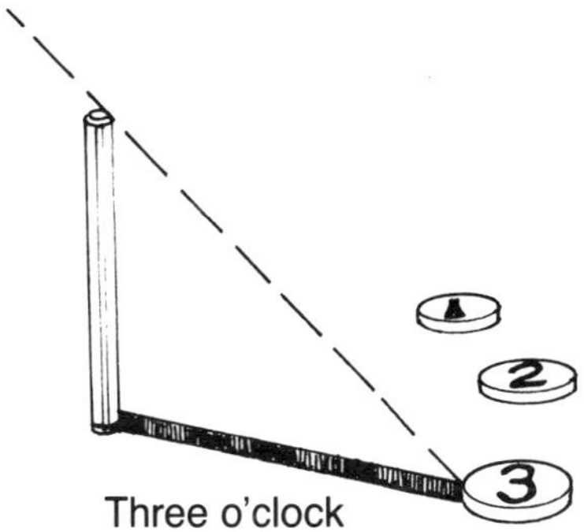

Three o'clock

There are many different types of sundials. Some are quite ancient, while others are fairly modern. The ancient Greeks, for example, used an **armillary** sundial. It was made of a bronze ring, several feet in diameter, which was mounted and set on a plane parallel to the **latitude** of the place where it was being used.

Compass sundials were introduced into Europe during the 1400s. Sundials have also been designed in the form of flat dials, cubed dials and hollowed-out globes.

ARMILLARY

An armillary is a model sphere with solid rings that are all circles of the sphere. These rings are used to represent the relationships between the heavenly bodies.

LATITUDE

The distance north and south of the equator that is measured in degrees on a globe or map.

An illustration of an armillary sphere from *The Iconographic Encyclopedia of Science, Literature and Art,* published in 1851.

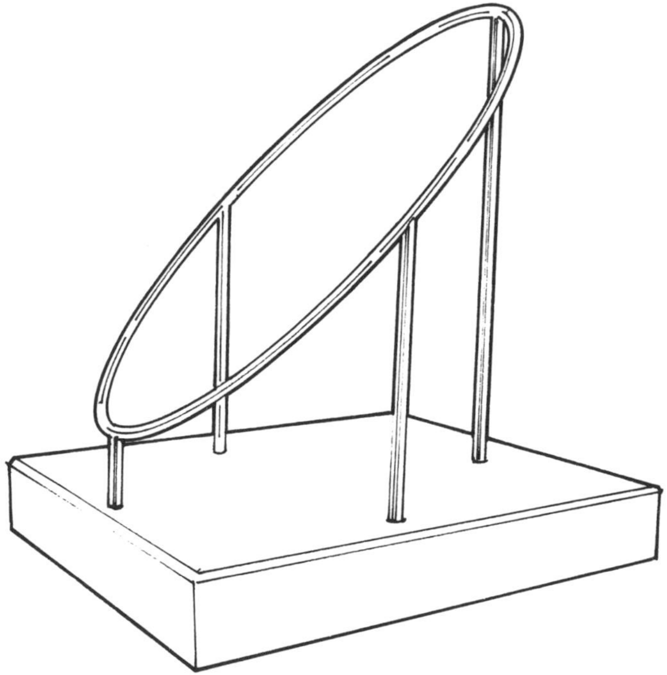

Greek Armillary Sundial

Basically all of these different types of sundials work in the same way. Each has a gnomon which casts a shadow on a numbered scale showing the hour. In order for the shadow to move uniformly over the face of the sundial, the gnomon must be set parallel to the Earth's north/south axis—in other words, due north above the equator and due south below the equator. The gnomon post is only set vertical at the North and South Poles. At other locations, it is set at an angle to the horizontal equal to the latitude of the location.

A sundial whose gnomon is set on a north/south axis, and whose horizontal angle is equal to the latitude where the sundial is set up, has many advantages. As the Earth rotates, the sun seems to make a circle around the gnomon every 24 hours. In fact, the Earth has rotated in relationship to the sun and the sun has remained constant. It is the shadow cast by the gnomon that is moving and marking off the hours, as the Earth is rotating in its daily orbit.

You can make your own sundial, which is adjusted for the latitude of the place where you live, by completing the following instructions. Do not get discouraged by the length of the directions. Even though they are long and require a lot of measuring, they are not difficult to do at all.

The first step is to collect the following materials:

Felt-tip pen
Protractor
Compass
Pencil
Ruler
Scissors or an X-Acto knife
White glue
2 pieces of cardboard, 8″ x 11″
Several sheets of 8″ x 11″ notebook paper

Before you can begin to make your own sundial, you must find out the latitude of the place where you live. You can do this by looking on a large-scale map. If you do not have a map that tells you the latitude, ask a reference librarian at your local library to help you find a map marked with latitudes. The weatherman at a local radio station or television station might also be able to tell you the latitude of your town.

Once you know the latitude of your town, complete the following steps:

1. Draw 2 lines, AB and CO, on a piece of cardboard so that they are perpendicular, or at a right angle of 90 degrees to one another. Line AB will be the hour line for six o'clock and line CO will be the hour line for twelve o'clock on your sundial.

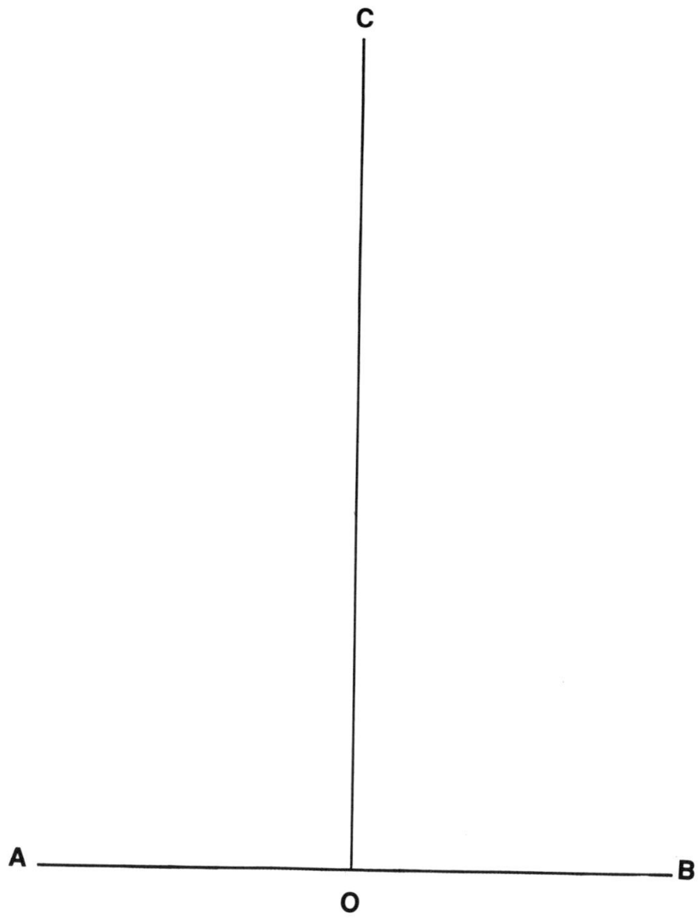

2. With your ruler and protractor draw line OD so that angle COD is equal to the latitude of where you live. The angle in the illustration below was drawn for a latitude of 40 degrees.

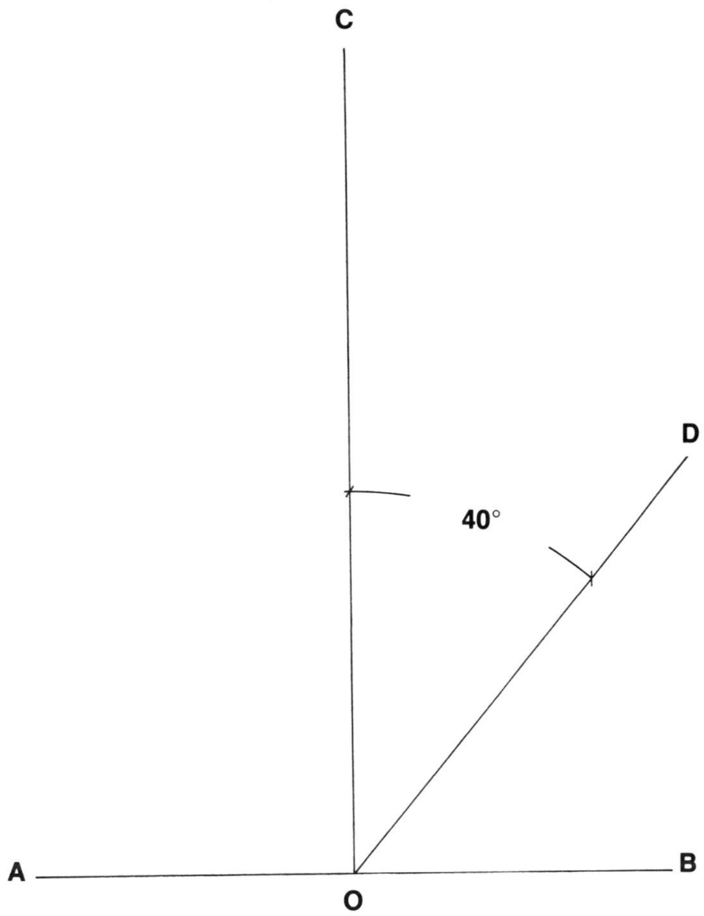

3. At any point along line OD draw a line EF that is perpendicular, or at a right angle of 90 degrees, to line OD.

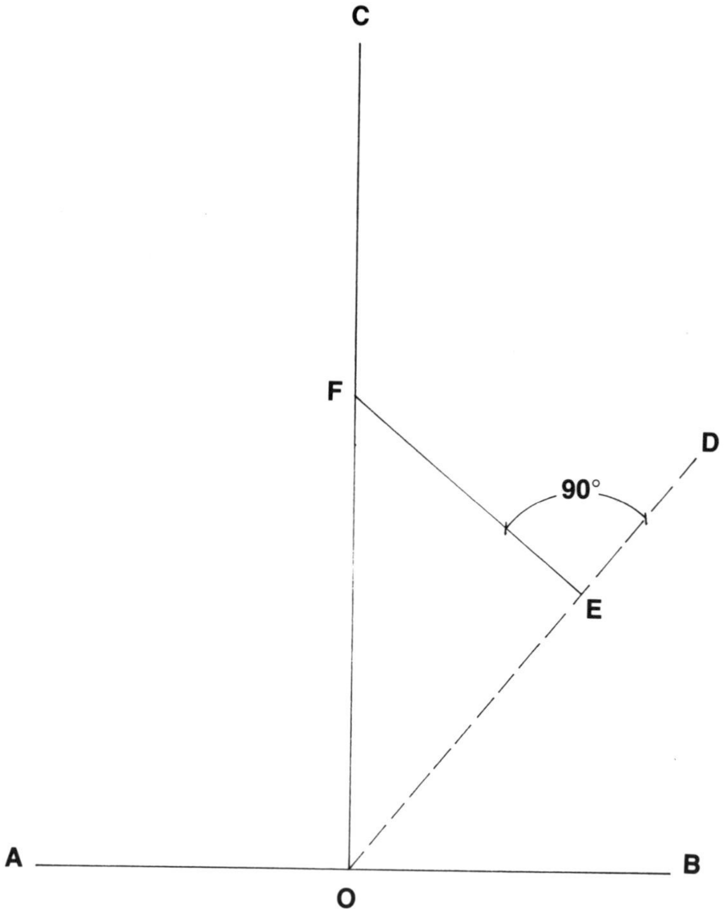

4. Mark off a distance on line CO, beginning at point F, that is equal in length to line EF. Label this new line, FG.

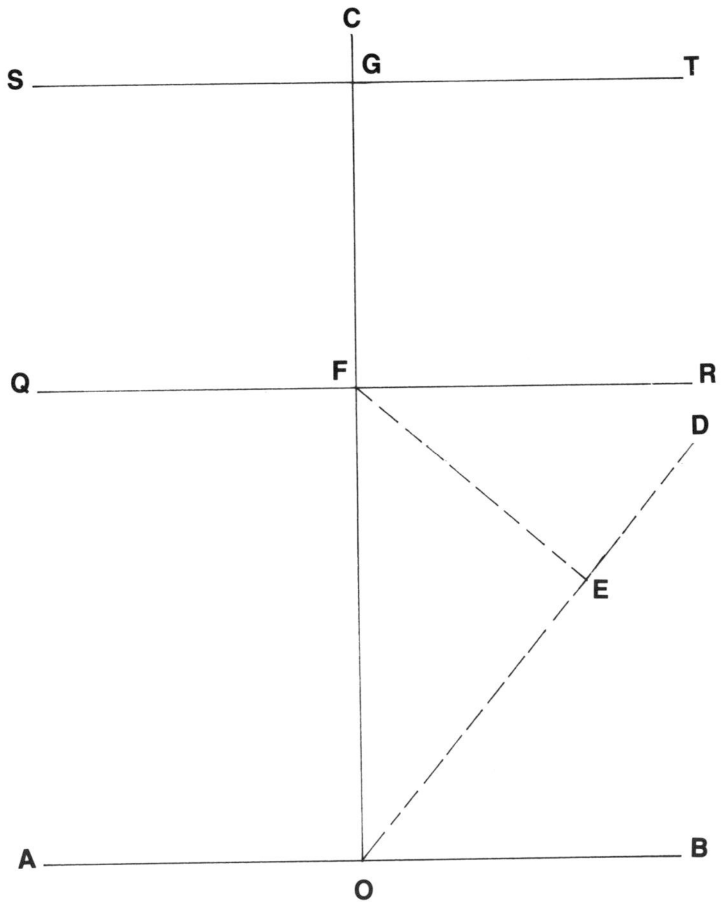

Draw lines SGT and QFR so that they are parallel to line AB.

5. Draw an arc with your compass from point T to point S, using point G as the center. Use your protractor to measure off 3 equal arcs, in each direction from point G, that measure 15 degrees each. Draw lines from point G to the marks for each of these arcs on line QR. Mark the intersections of these lines with line QR from left to right: 9, 10, 11, 12, 1, 2, and 3 as illustrated.

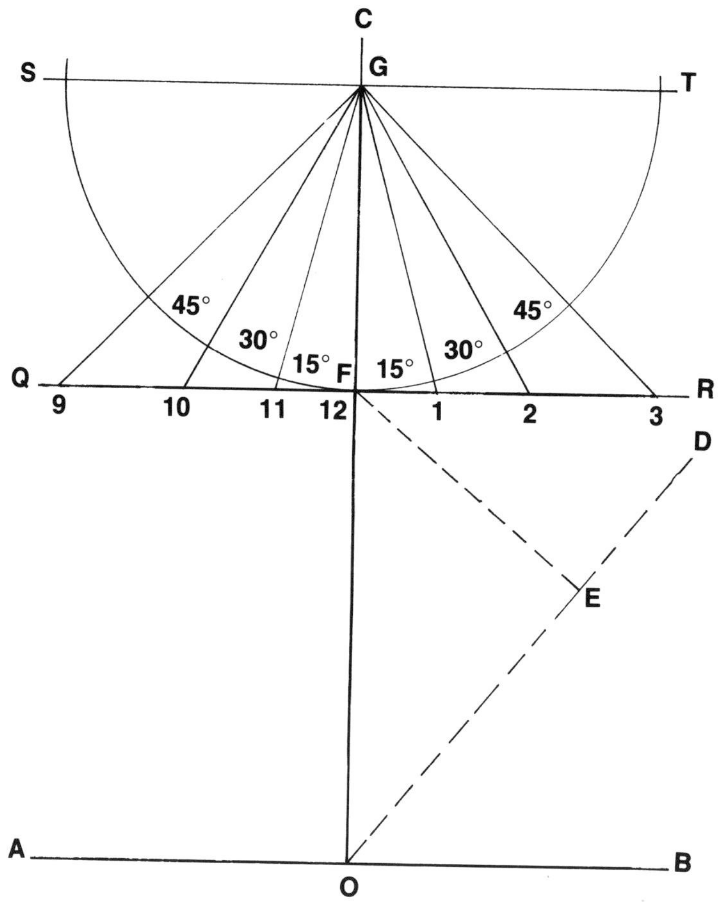

6. Draw lines from point O to the 7 points you just marked with numerals on line QR. These lines will be the hour lines for your sundial for nine o'clock, ten o'clock, and so forth.

Draw line FW so that it is parallel to the nine o'clock line that extends from the point 9 to point O. You can do this by measuring the distance from point 9 to point F. Point W should be the same distance from point O as the distance you just measured. Connect point W with point F.

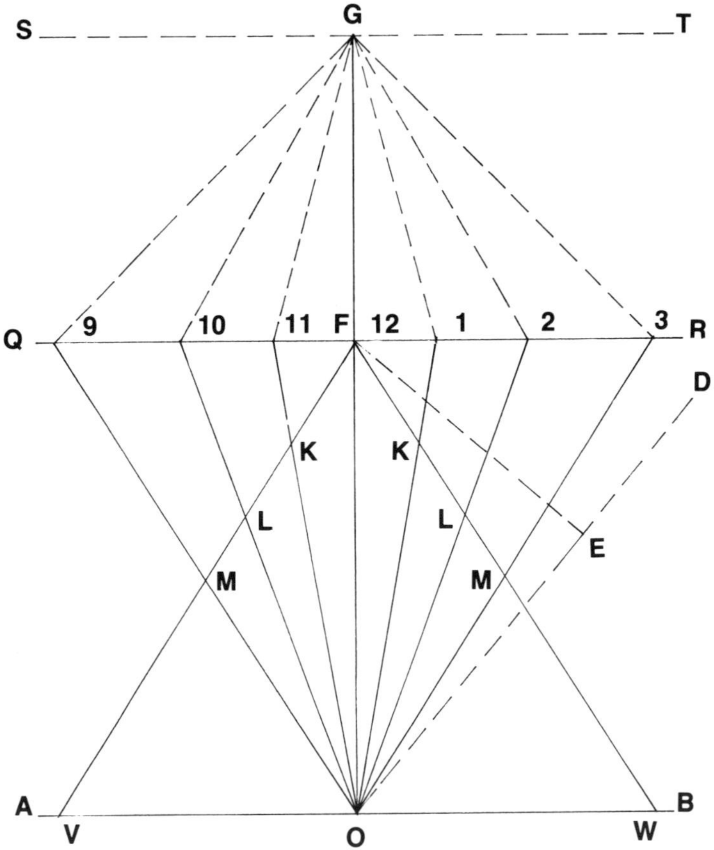

Then measure the same distance on line *AB* from point *O* to a new point *V*, in the direciton of OA. Connect point *V* with point *F* with a line that will be parallel to line *O3*. Mark the points of intersection on lines *VF* and *WF* as *M, L, K,* and *K, L, M* as illustrated.

7. Place the point of your compass on point *M* on line *FW*. Draw a 180 degree arc from point *L* until it intersects line *FW*. Mark the point of intersection as *N*.

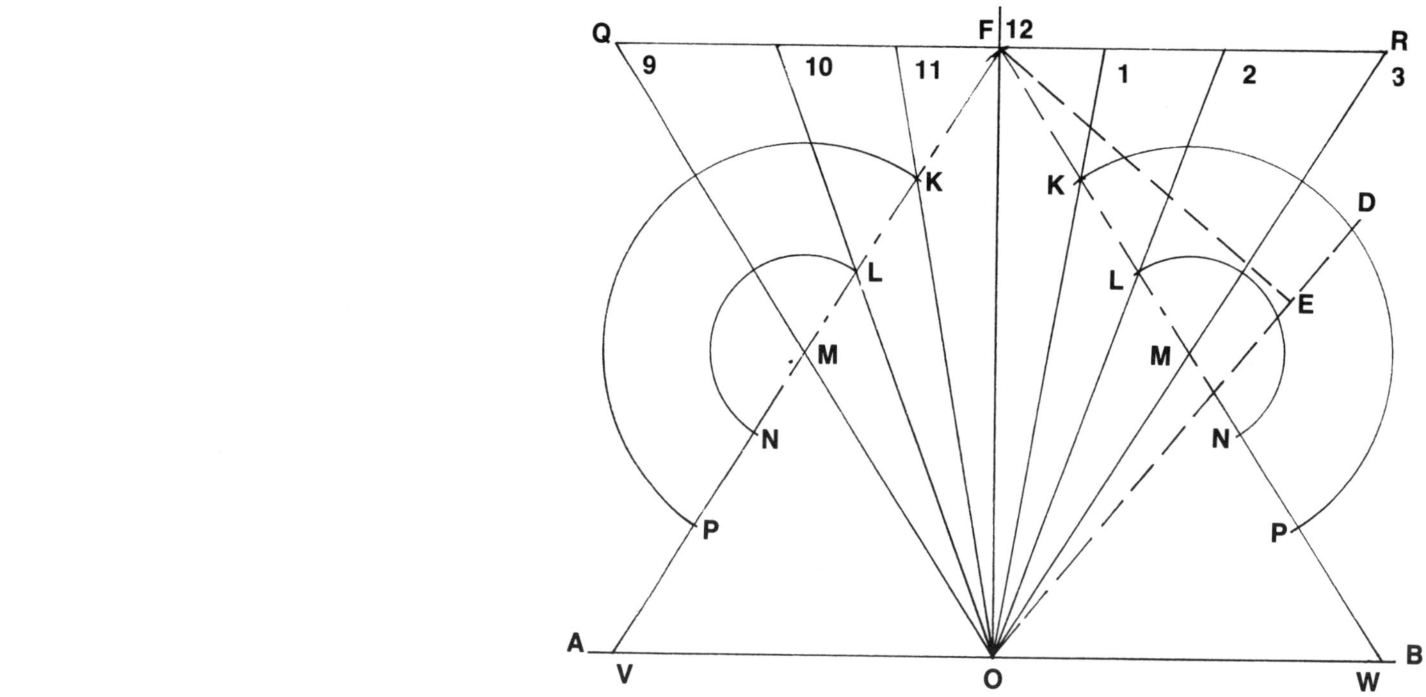

Now place the point of your compass on point *M* again and draw another 180 degree arc from point *K* on line *FW* until it intersects line *FW* again. Mark this point *P*.

Repeat the same steps by drawing two 180 degree arcs on line *FV* as illustrated.

8. Draw a line to connect point Q with V, and point R with W. Draw a line from point O through point P to line RW. Mark the point of intersection as 5. Draw a line from point O through point N to line RW and mark that point of intersection as 4. These are the four o'clock and five o'clock hour marks for your sundial.

Repeat the same steps from point O through points P and N on line FV through to line QV. Mark these points of intersection, 7 and 8, as illustrated. These are the seven o'clock and eight o'clock hour marks for your sundial.

Mark point W on line segment OB with a 6 for the 6:00 P.M. hour mark for your sundial and point V on line segment OA for the 6:00 A.M. hour mark.

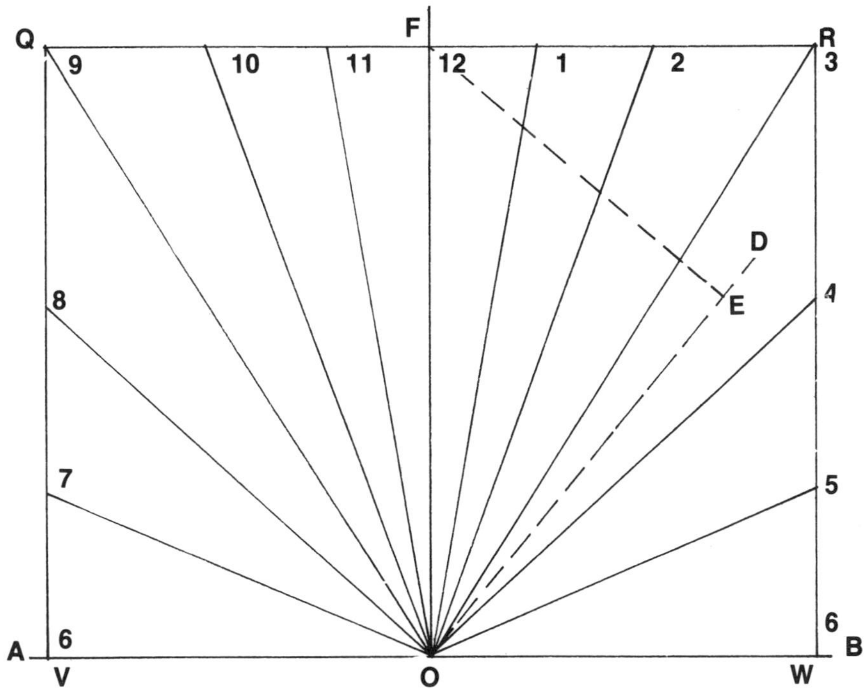

9. In order to derive the hour line for 4:00 A.M., extend the line segment for 4:00 P.M. in a straight line on through point O as illustrated. Do the same with the line for 5:00 P.M. to obtain the point for 5:00 A.M.; 7:00 P.M. for 7:00 A.M.; and 8:00 P.M. for 8:00 A.M. Extend lines RB and QA to connect these points and label the points appropriately as illustrated. The dial plate for your sundial is now complete.

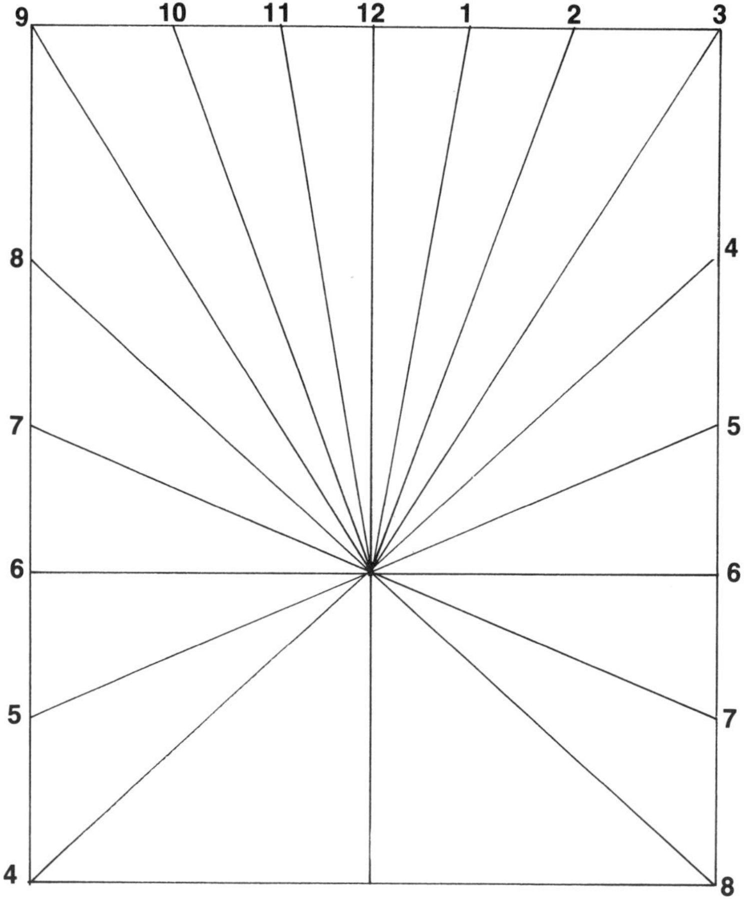

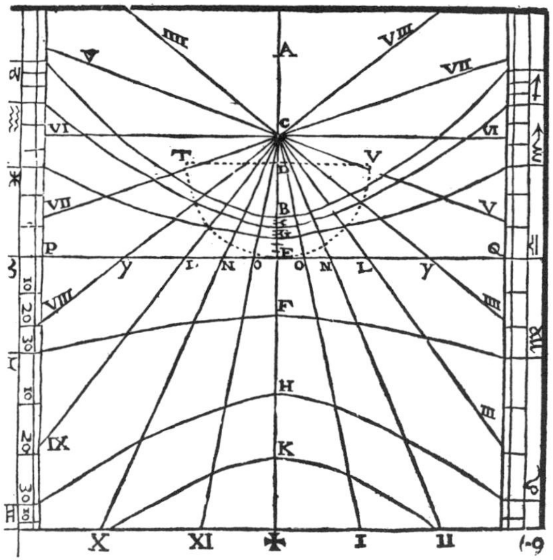

A sundial plate illustrated in John Blagrave's *The Art of Dyalling in Two Parts,* published in 1609.

10. The final step in making your own sundial is to make the gnomon. On a separate piece of paper draw 2 lines, AB and CO, that are perpendicular to one another.

Using your protractor, measure an angle equal in degrees to the latitude where you live and will be using your sundial. The center of the angle should be at point O. One side of the angle should be the line CO. Draw line DO so that it forms the other side of the angle.

Connect lines DO and CO with the fancy curved line FE as illustrated. This curve is simply to decorate your gnomon.

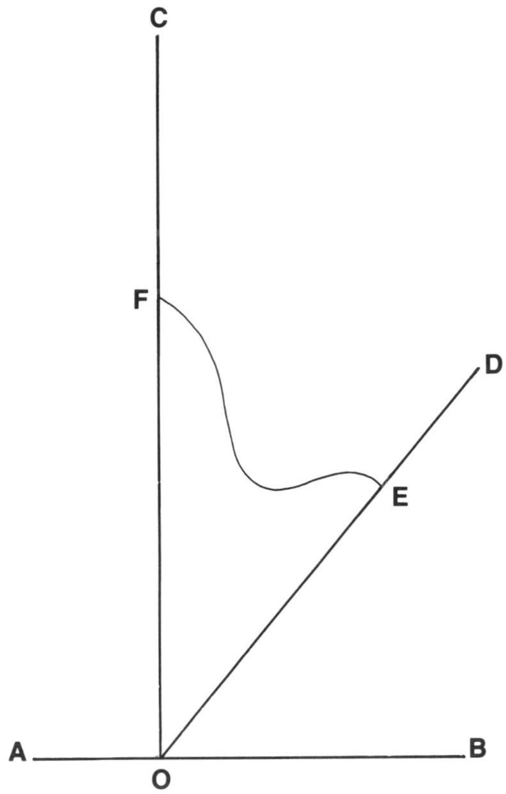

11. Cut out the pattern for your dial plate and mount it on a piece of cardboard. Trim the pattern with an X-Acto knife or scissors.

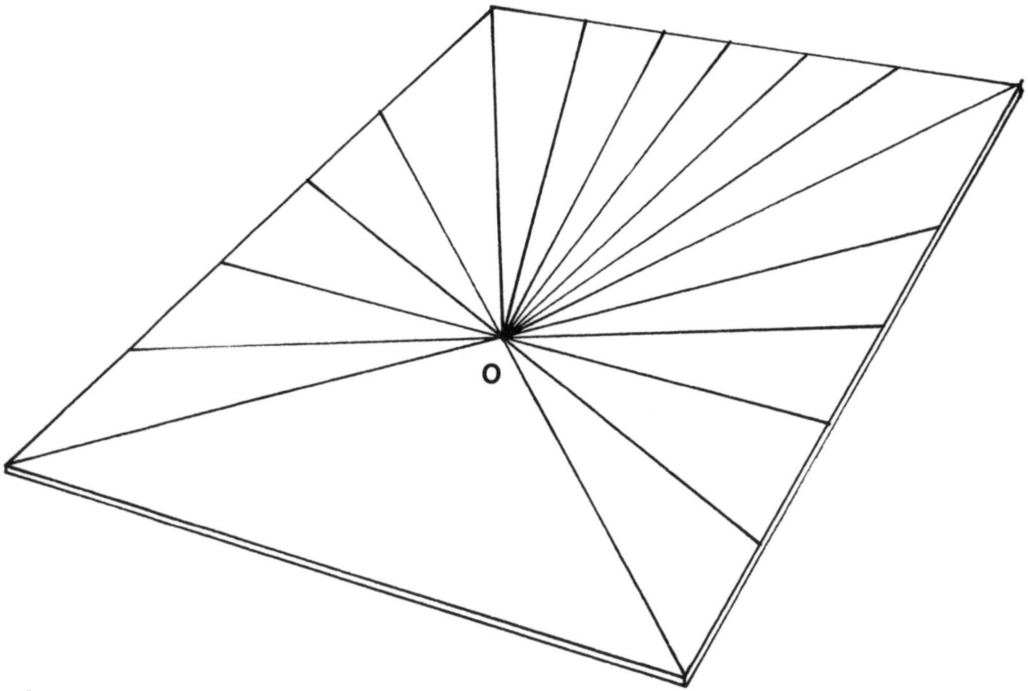

Cut out the pattern for the gnomon and also mount it on a piece of cardboard. Trim the edges with the X-Acto knife or scissors.

12. Now you are ready to assemble your sundial. Apply glue along the base of the gnomon (line OE) and attach it to the dial plate at point O, along the twelve o'clock hour line, as illustrated.

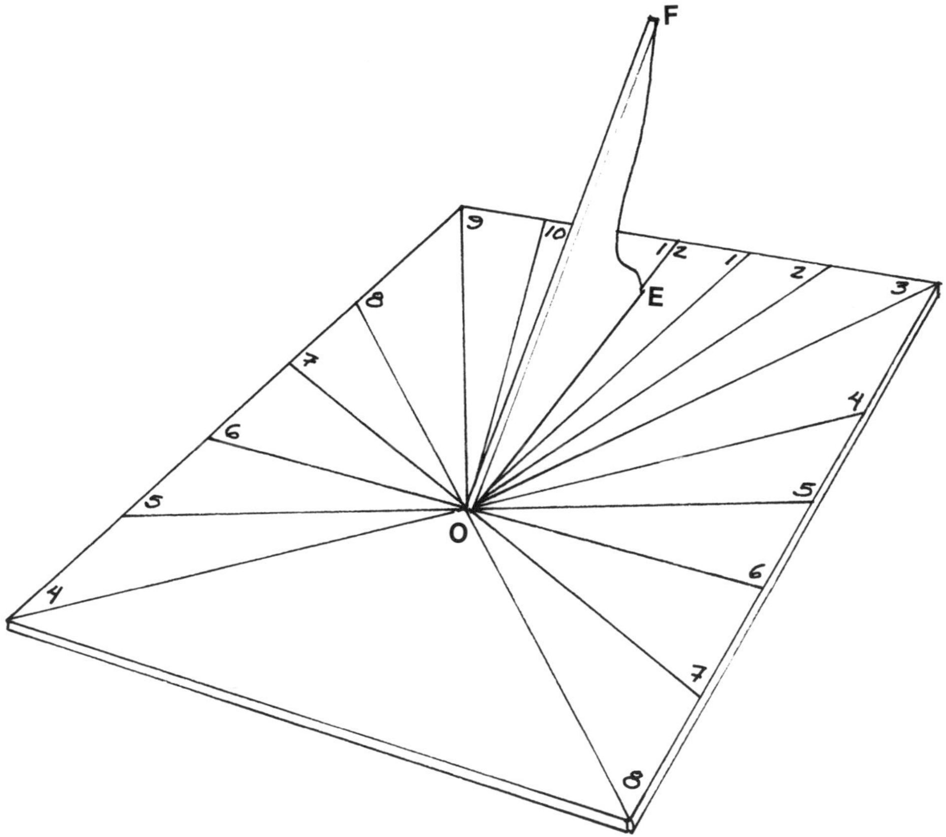

Your sundial is now complete and ready to use. Place it on a level base, with the 12:00 A.M. hour mark facing due north. On sunny days you can forget your watch and tell time the way the ancients did, and people have for hundreds of years since, by using the sun.

A sundial from the early 1700s

Sundials can be found all over the world. Their shape and construction may tell you a lot about their builder's knowledge of mathematics and astronomy. Some early dials were inscribed in the stone of churches to note the times of prayer. Others might be etched in metal, or carved in wood, and mounted on the side of a building. They may also be found carved in stone and mounted on pedestals in parks, gardens and cemeteries. Sometimes these dials are so elaborate that you can hardly read the time of day on them. Others are inscribed with mottos like this: Behold and begone about your business. Sundial verses are found all over the world, usually in gardens and parks. They are meant to be read when the time is read, like this one from the 1700s:

A riddle I will put to thee:
Who gives thee years yet takes of thine?
Who lends thee most when he robs thee?
You've guessed, no doubt—'tis Father Time.

The ancient Greeks made extensive use of sundials. They were probably introduced into Greece by the astronomer, **Anaximander**, around 600 B.C. He may have learned about the sundial and gnomon from the Babylonians.

By the end of the 500s B.C., Babylonian astronomers had divided the heavens into 12 constellations (clusters of stars), 30 degrees apart, in a circle of 360 degrees. This system became the basis for the **zodiac**, still used by astrologers today, and was also the foundation of our 12-hour day and night.

While Babylonian astronomers had to rely upon systematic observations of the solar system, the Greeks had enough knowledge of

ANAXIMANDER

A Greek astronomer and philosopher who lived from 611 to 547 B.C.

geometry to attempt to understand and explain it. Like most ancient people, the earliest Greeks had attempted to explain the cosmos through myths of gods and goddesses who were responsible for creating the Earth and man. It was Anaximander who first tried to describe the planets and stars in a scientific way. His explanation put the Earth in the center of the universe. Since it was exactly in the center, it had no reason to fall one way or the other. But Anaximander did not try to explain how the planets moved.

This explanation was left to **Pythagoras** and his students, who argued, 50 years later, that the planets all moved around the Earth in circular orbits. They also believed that the universe itself was a sphere, and that the sun and moon traveled around the Earth in circular orbits. These Greek philosophers believed the order of the universe could be explained according to the laws of geometry.

Other Greeks, however, came up with new ideas about the universe. One of the most important was that of Herakleides, who lived during the 300s B.C. He believed that the Earth rotated on its **axis** once a day. Herakleides also believed that Mercury and Venus revolved around the sun, which itself revolved around the Earth. Another Greek, Aristarchus, who was born in 310 B.C. (the year Herakleides died), carried this new notion of how the solar system worked one step further. Aristarchus described a system with the sun at the center. The other planets, including Earth, revolved around the sun in orbits of varying distances from the sun. Only the moon still orbited the Earth.

Unfortunately these theories, which were very close to being correct, were not widely accepted and were eventually lost and buried for over 1,000 years. Instead, the last great Greek astronomer, **Ptolemy**, was responsible for a mistaken idea of how the universe was structured. His theory was popular for hundreds of years. Ptolemy lived and worked in Alexandria, Egypt, which was the center of world

The signs of the zodiac

ZODIAC
An imaginary belt of the heavens that extends on both sides of the path of the sun. The planets and the moon are included in this belt. The zodiac is divided into 12 equal parts named after 12 groups of stars known as the signs of the zodiac.

GEOMETRY
The mathematics of space, which studies the relationships of points, lines, angles and surfaces of figures in space.

PYTHAGORAS
Greek philosopher who lived during the 500s B.C.

AXIS
The axis of the Earth is an imaginary line through the North Pole and the South Pole about which the Earth turns.

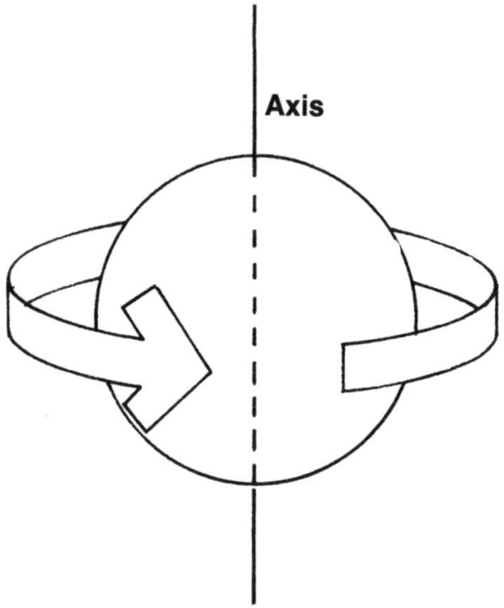

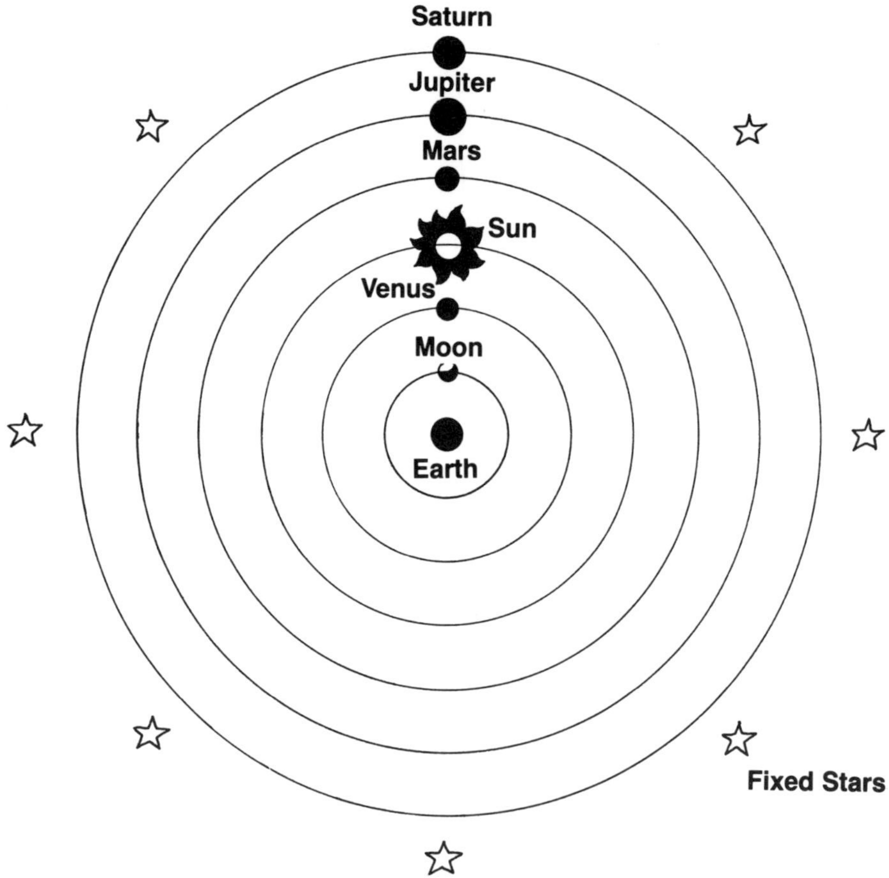

The solar system according to Ptolemy

PTOLEMY

Greek astronomer, mathematician and geographer who lived during the A.D. 100s. Ptolemy is pronounced tol ′e-me.

ARISTOTLE

Greek philosopher who wrote works on philosophy, natural science, politics, ethics and poetics. He lived from 384 to 322 B.C. Aristotle also was the tutor of Alexander the Great.

EPICYCLES

In Ptolemaic cosmology, a small circle of a heavenly body, the center of which moves around the Earth on the deferent, or circumference of an orbit, whose center is the Earth.

trade and science. He revived the idea of **Aristotle** that the Earth is the center of the cosmos and that God is responsible for the movement of the other heavenly bodies. The Earth remained the unmovable center of the universe in Ptolemy's system. The moon, the sun and the other stars traveled in small, circular orbits, or **epicycles**, around the Earth on deferents, or larger circular orbits. The epicycle theory of Ptolemy explained why some planets occasionally appeared to be moving unevenly in their circular orbits around the Earth.

Ptolemy was also responsible for a star table, which listed 1,022 stars and their positions in the sky above Alexandria. He described the construction of several instruments, including the astrolabe, which could be used for measuring the positions of the stars. The astrolabe was to become the basic tool for astronomers for the next 1,000 years. By the Middle Ages, elaborate versions of the instrument were developed, which were used not only to study the stars, but also to tell time.

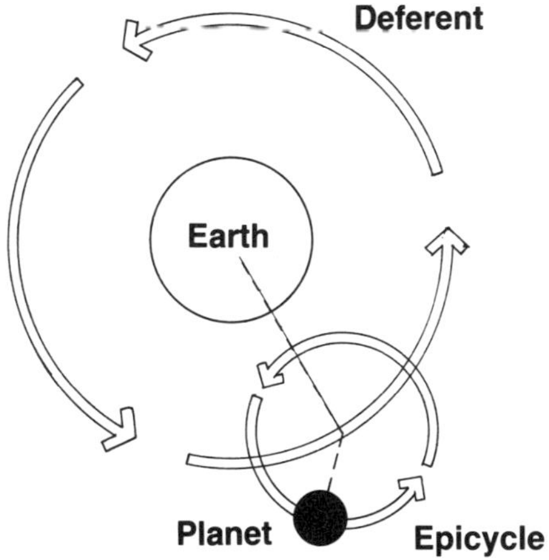

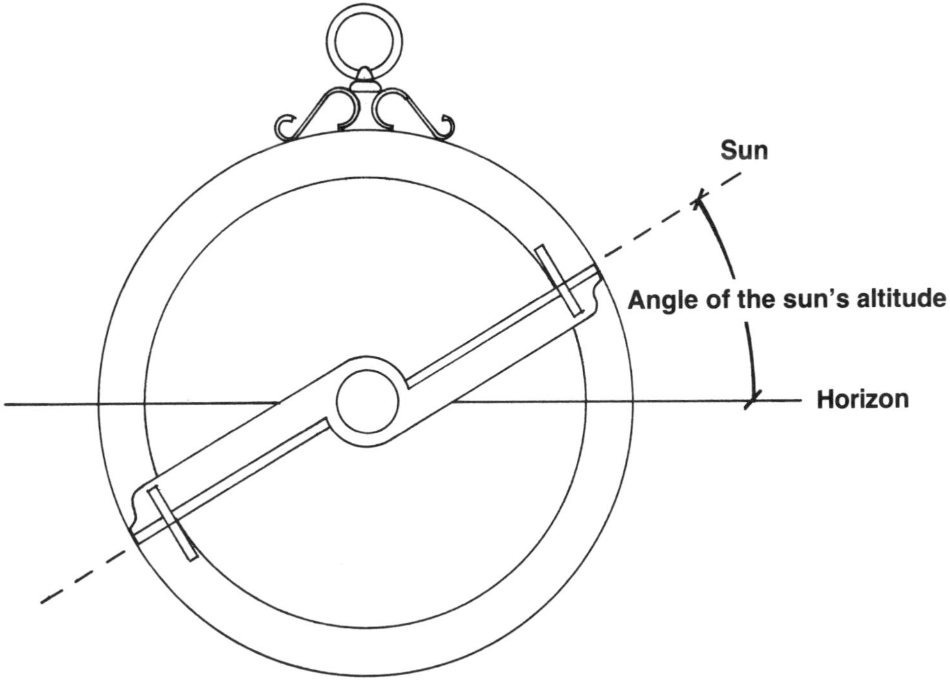

Diagram of how a simple astrolabe works.

ALTITUDE

The angle from the horizon line up to a star or heavenly body.

In its simplest form, the astrolabe consisted of a pivoted arm with sighting holes at the upper and lower ends. It was used to determine the **altitude** of the sun and other stars. When the instrument was hung from its ring, the pivoted arm was moved until a beam of sunlight passed through the sighting hole on the upper end of the arm through to the sighting hole on the lower end. The angle between the upper arm and the horizon was the altitude of the sun above the horizon. At night a sighting was made directly through the two holes to determine the altitude of a star.

An illustration of Ptolemy observing the stars from the Italian edition of his work, *La Geographia,* printed in 1547.

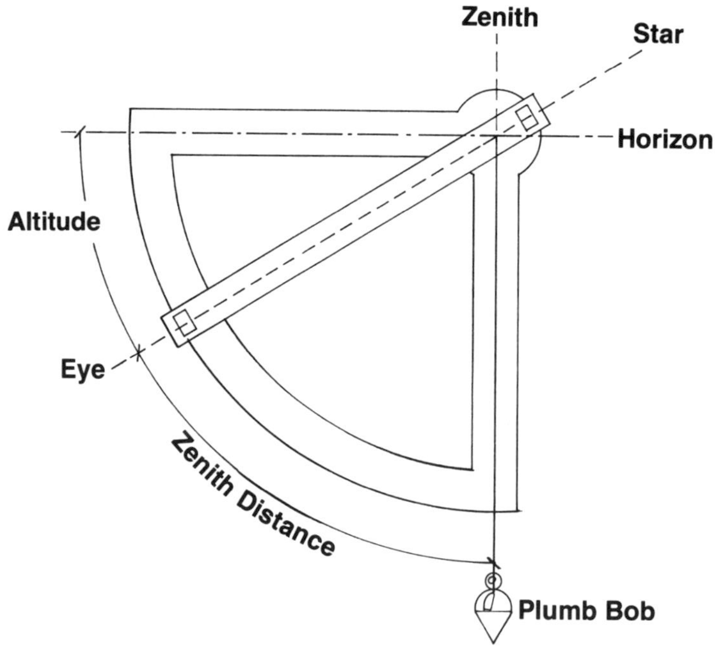

Diagram of a quadrant

Another device believed to have been invented by Ptolemy was the quadrant. A quadrant is a quarter of a circle with the angles (degrees) marked on its edge. The instrument was adjusted to lie on the place of the meridian, or horizon, where it was being used. A plumb bob, suspended from its center, established its vertical alignment with the **zenith**. A movable rod, about the same length as the radius of the quadrant, was pivoted from the center. Sighting mounts were attached to this rod. When you observed the stars and planets through the quadrant by pivoting it, the altitude of the star would be indicated by the plumb line on the quadrant's degree scale.

The Arabs modified the quadrant by developing a means by which it could rotate around a vertical axis. This improvement was extremely important, since it not only made it possible to measure an object's altitude above the horizon, but also its angular distance or **azimuth** from a fixed point on the horizon. The altitude and azimuth are the basic astronomical measurements of a heavenly body or star.

ZENITH

The point in the sky or celestial sphere directly overhead of the observer.

AZIMUTH

The angle measured around the horizon from the north point to a point under a star which is being located.

An illustration of Arabian astronomers scanning the heavens from Macrobius' *In Somnium Scipionis,* (1513).

An illustration of how an astrolabe is used to measure altitudes. From a 1674 book on astronomy and geography.

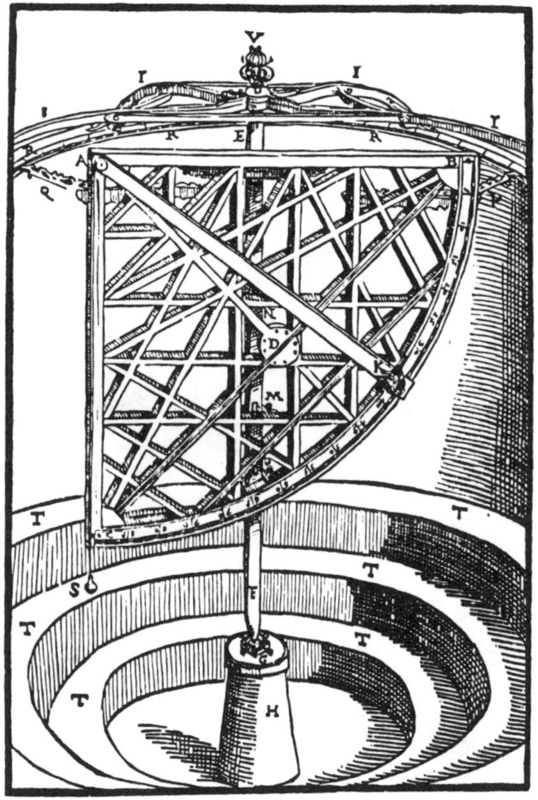

An illustration, from the 1500s, of a revolving azimuth quadrant.—Tycho Brahe, *Astronomiae Instauratae Mechanica*

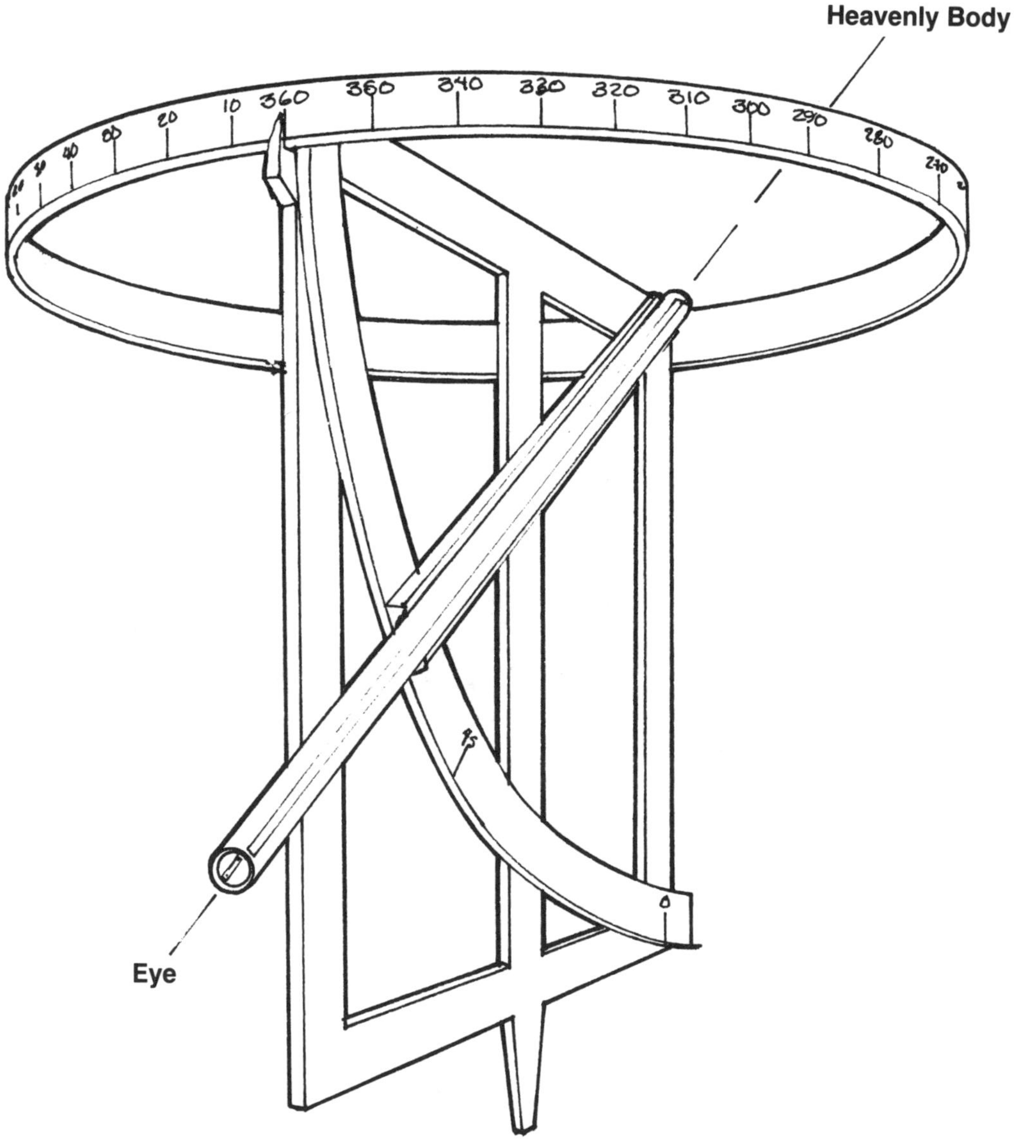

A quadrant is used to determine the star's altitude above the horizon and its azimuth from the north-south meridian.

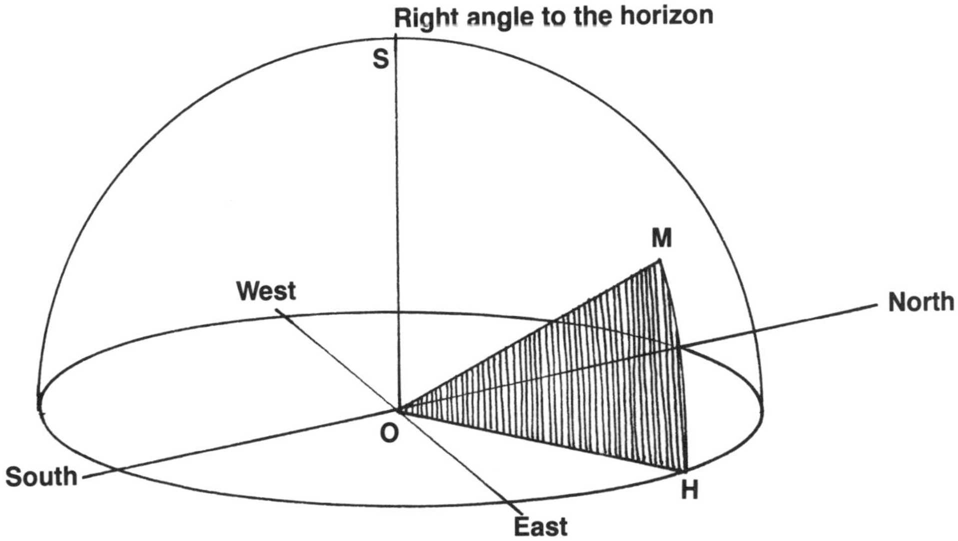

The angle MOH is the angle for the altitude for the object, or star M. SOH is the angle of the object's azimuth.

You can make a quadrant, that will rotate around a vertical axis, just like an Arabian azimuth quadrant. You will need the following materials:

- Piece of 1/16″ thick cardboard, 8″ x 11″ (The back of a tablet will work very well.)
- Piece of lightweight cardboard, 3″ x 2″ (A file folder will work well.)
- 8″ x 7″ piece of 1/16″-thick masonite
- Tracing paper or thin white paper
- Thread spool, plastic or wooden, 1¾″ tall
- 12″ piece of medium-weight thread
- 2 pushpins or straight pins
- 1 plastic peashooter
- Three ¼•20 bolts with nuts
- A piece of doweling, ¼″ in diameter and at least 8″ long

- Newspapers
- Black spray paint
- Spray mount
- White glue
- Epoxy glue
- Small saw
- Hand drill
- X-Acto knife
- Optional: Bubble, or carpenter's level

Copy the quadrant patterns on pages 51-53 onto tracing paper. The patterns can also be photocopied. Mount the patterns on the pieces of cardboard and cut them out with an X-Acto knife.

Spread out the newspapers on the floor. Spray paint the thread spool, the 7″-long piece of doweling and the peashooter. Let them dry thoroughly.

Mount the pattern for the quadrant base on page 00 onto the piece of masonite or wood. Drill four ¼″ holes in each of the points indicated on the pattern.

Glue the nuts from the 3 bolts onto the bottom of the quadrant base with the epoxy glue, as illustrated.

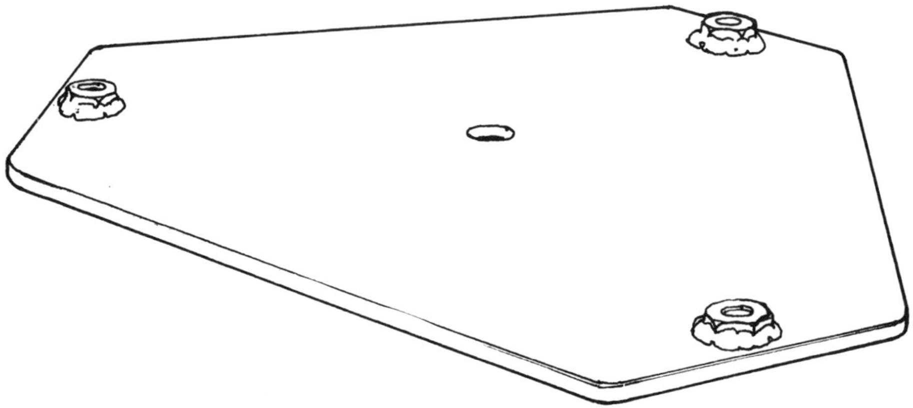

Glue the ¾″-long piece of doweling into the center hole on the base of the quadrant. When the nuts have dried on the bottom of the quadrant base, screw in the bolts as shown below.

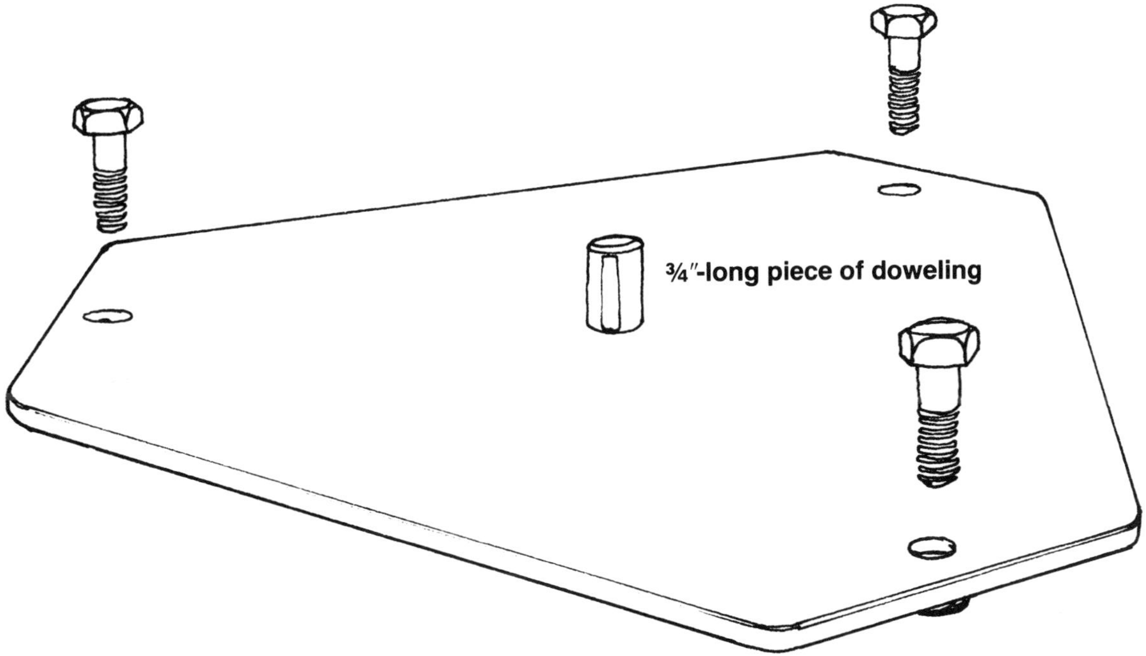

Cut a hole slightly larger than ¼″ in diameter in the center of the degree circle. Glue the degree circle to the top of the base, using the pattern lines to help center it.

Place some white glue on one end of the 7″ piece of doweling. Insert the dowel into either end of the thread spool. Glue the pointer to the other end of the spool so that the hole in the spool and the hole in the center of the pointer match up. Be sure that you have cut out the center hole in the pointer.

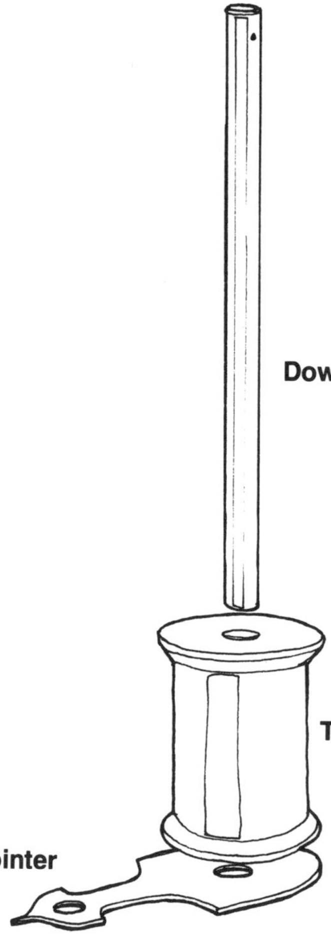

PROTRACTOR

An instrument for measuring or drawing angles.

Push the pushpin through the center of the plastic peashooter and the **protractor** piece of the quadrant as illustrated. If the peashooter is too long, trim its ends. One half inch should stick out beyond each end of the protractor.

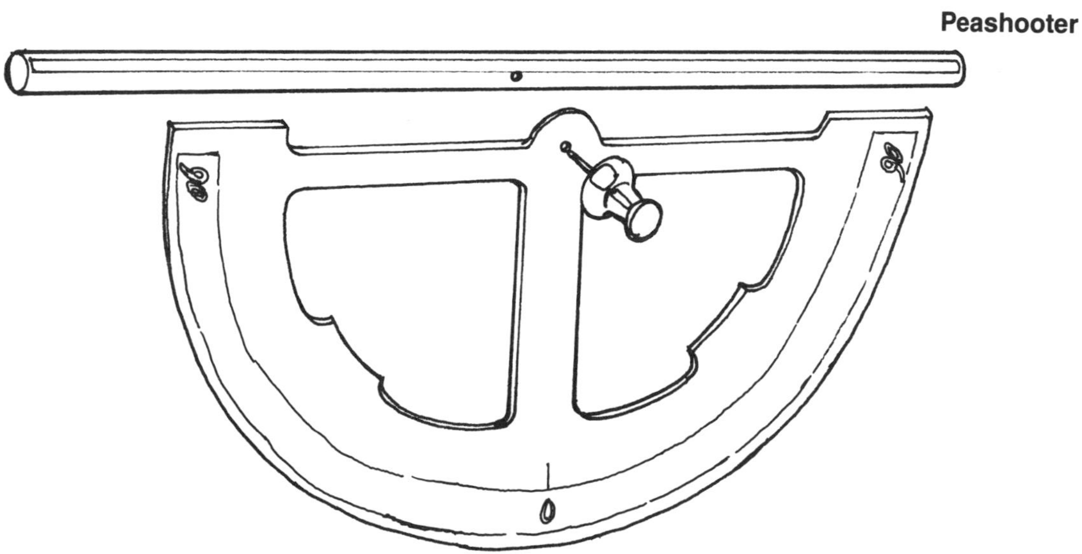

Glue the peashooter to the edge of the protractor so that it is aligned with the 90 degree marks at each end of the protractor. The marks for the 90 degree points on the protractor should be at the center of the peashooter.

Push the pushpin, attached to the protractor and peashooter, through the top of the 7″ piece of dowel, which is already glued to the thread spool and pointer. If a pushpin is not long enough to go through both the protractor and the peashooter, use a straight pin. Be sure to bend down the tip of the straight pin after you have pushed it through the piece of dowel. Loop and knot the 12″ piece of thread around the pushpin and attach the second pushpin to make a plumb bob as illustrated. The plumb line, created by the piece of thread and the plumb bob, is used to measure the degrees of altitude on the quadrant's protractor.

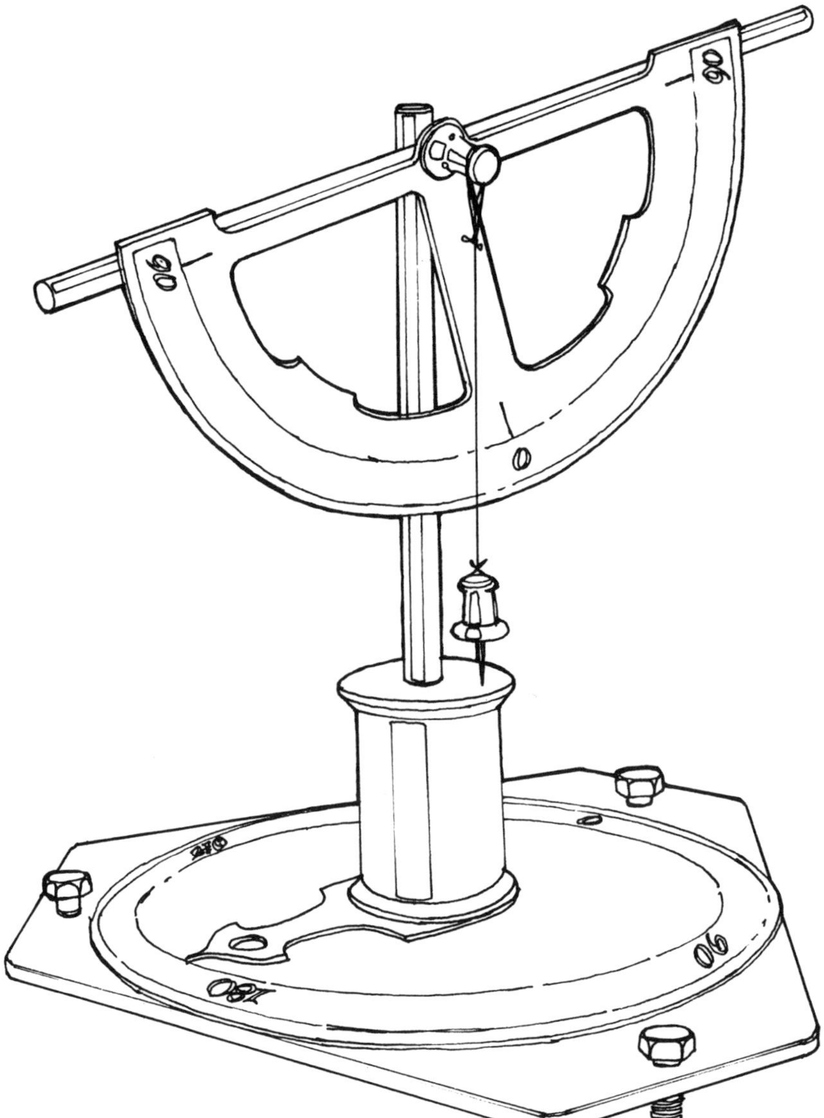

Protractor

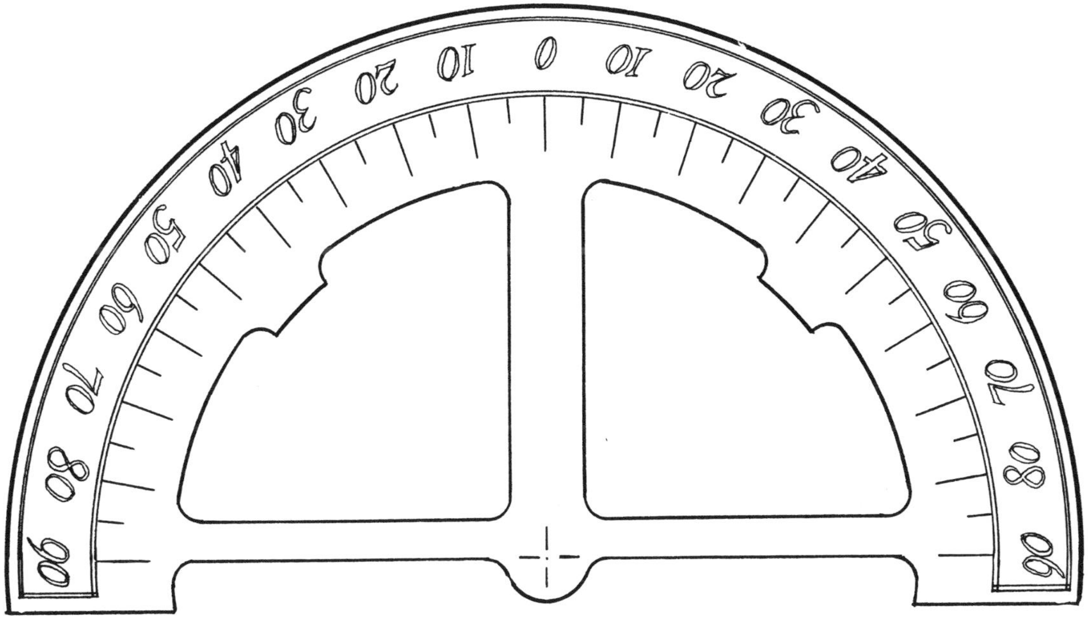

Degree Circle

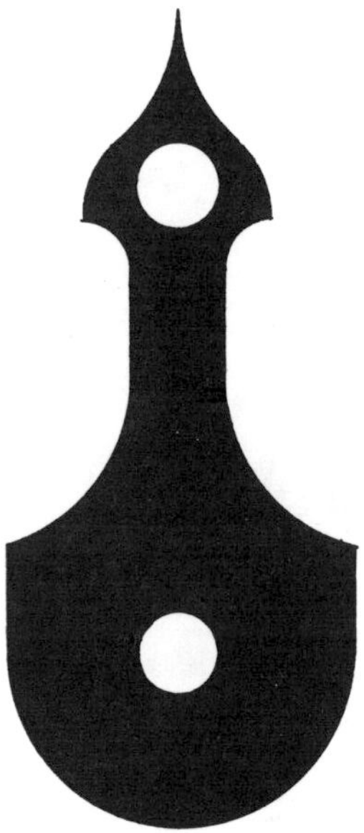

Pointer

QUADRANT BASE PATTERN

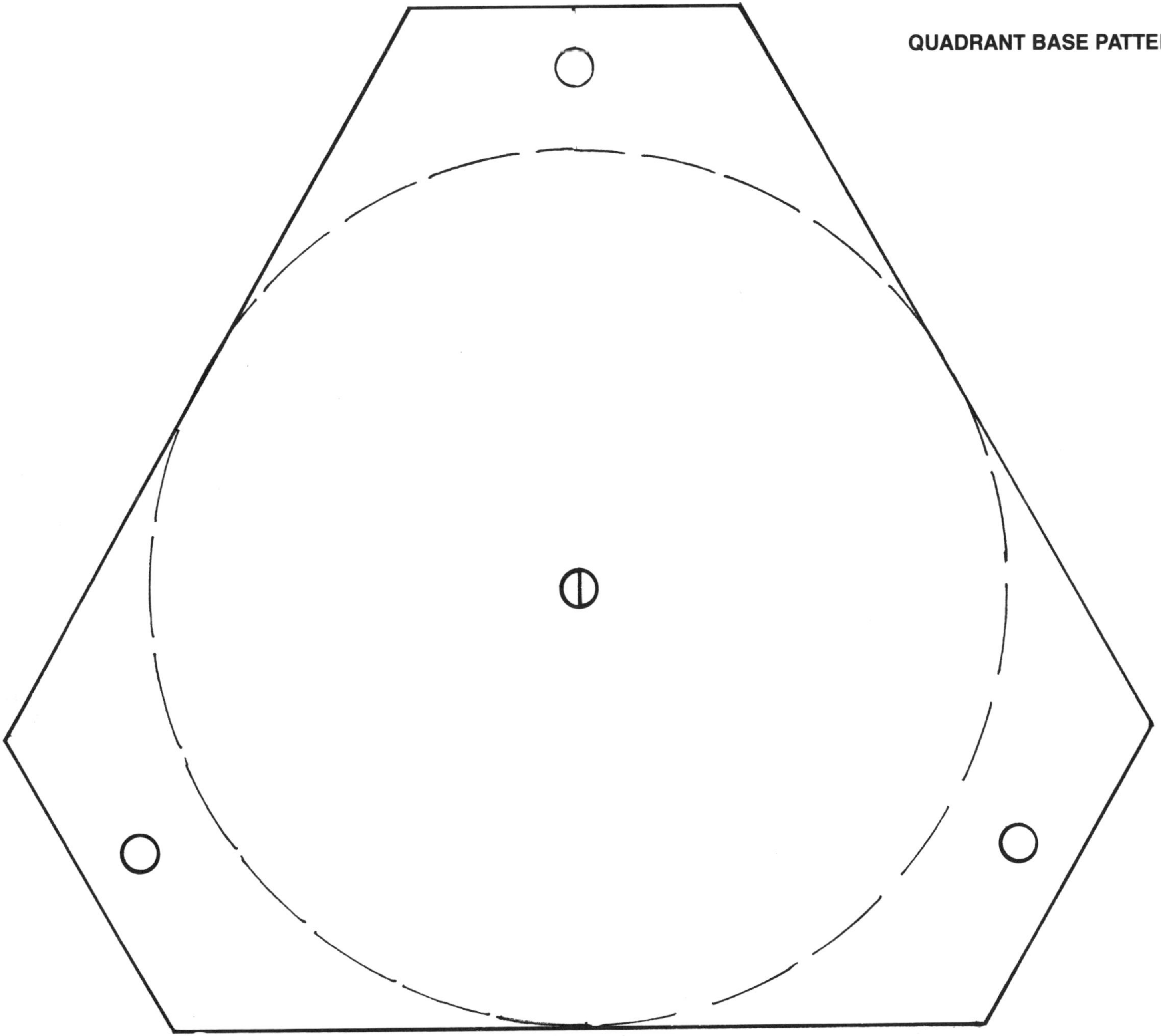

Your quadrant is now complete. To use it, set it on a flat surface. Adjust the screws in the base of the quadrant to make it level. You can check if your quadrant is level with a bubble, or carpenter's, level. Sight an object, such as the top of a tall building or tree, with the sighting rod. If it is night, you can observe a star.

The degree indicated on the base of the quadrant, by the pointer, is the azimuth measurement. The number of degrees indicated on the protractor by the plumb line will be the altitude measurement of the object you are plotting.

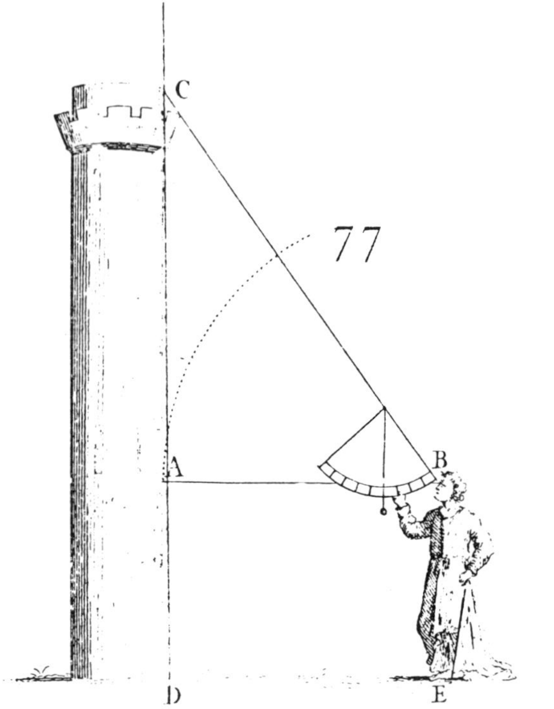

An illustration of a man using a quadrant to measure the height of a building. From Roger Long's *Astronomy in Five Books,* (1742).

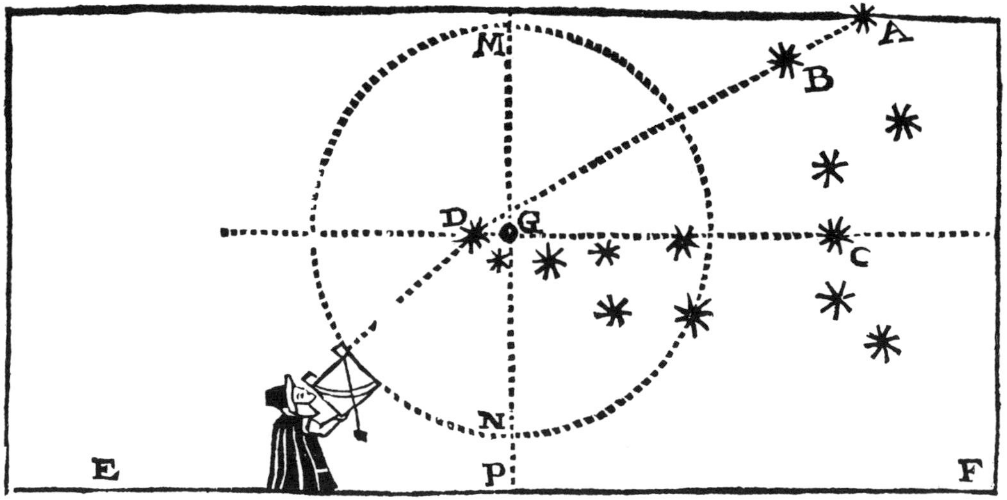

An astronomer plotting the stars with a non-revolving, or simple Ptolemaic, quadrant. From John Blagrave, *The Art of Dyalling in Two Parts,* (1609).

The quadrant is an instrument that can divide the sky overhead into a giant inverted bowl. When you make an observation with your quadrant it gives you the precise location of a star along a particular set of **axes.**

The circle around which your quadrant rotates is level with the horizon and is known as the azimuth. The azimuth is divided into 360 degrees of rotation, or a full circle.

AXIS (Plural—Axes)

Central line around which parts are regularly or evenly arranged.

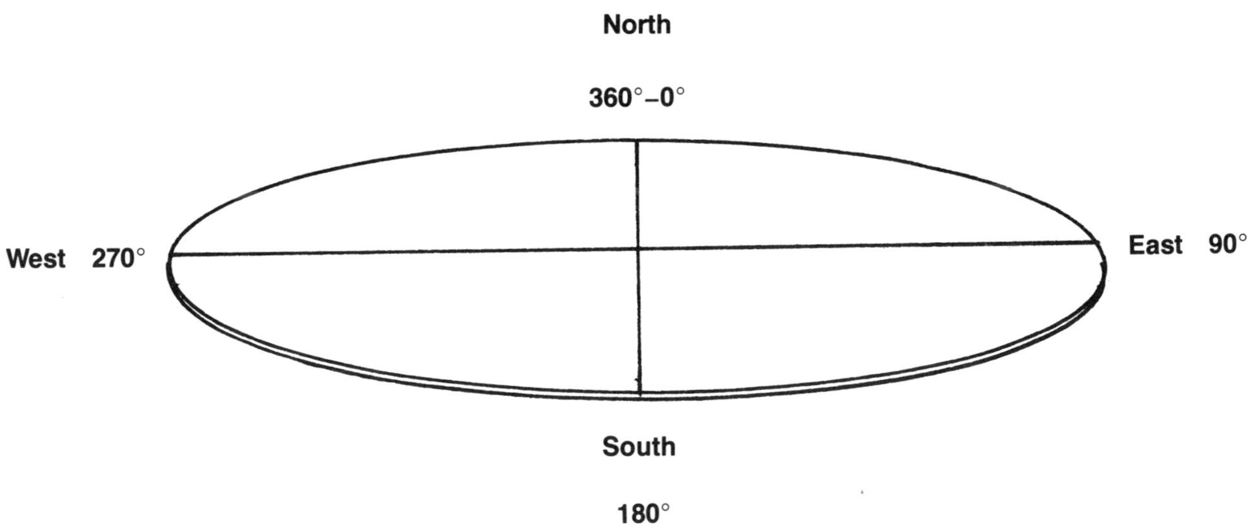

The half circle of 180 degrees that your quadrant measures overhead is used to measure the altitude of the star. Directly above you is 90 degrees, or the zenith, which is the highest point in the sky.

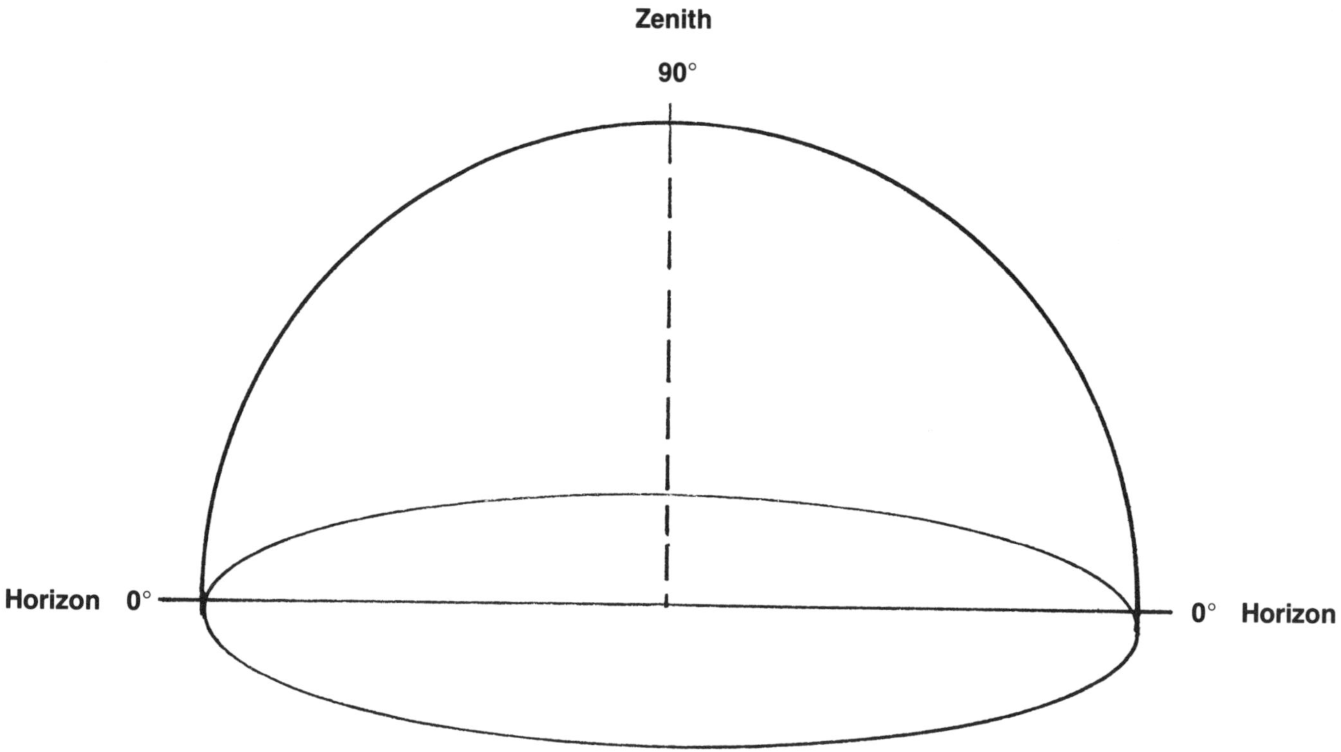

When you plot a star's location with your quadrant you will note its degree of azimuth and the degree of its altitude. In order to make a permanent record of the location of the stars you will observe, you may want to make your own star map. For hundreds of years, astronomers have used star maps to help them recognize stars and constellations in the zodiac.

Star map of stars and constellations seen in the Northern Hemisphere. The astronomer Ptolemy is in the upper right-hand corner. Engraved in the early 1500s by Albrecht Durer.

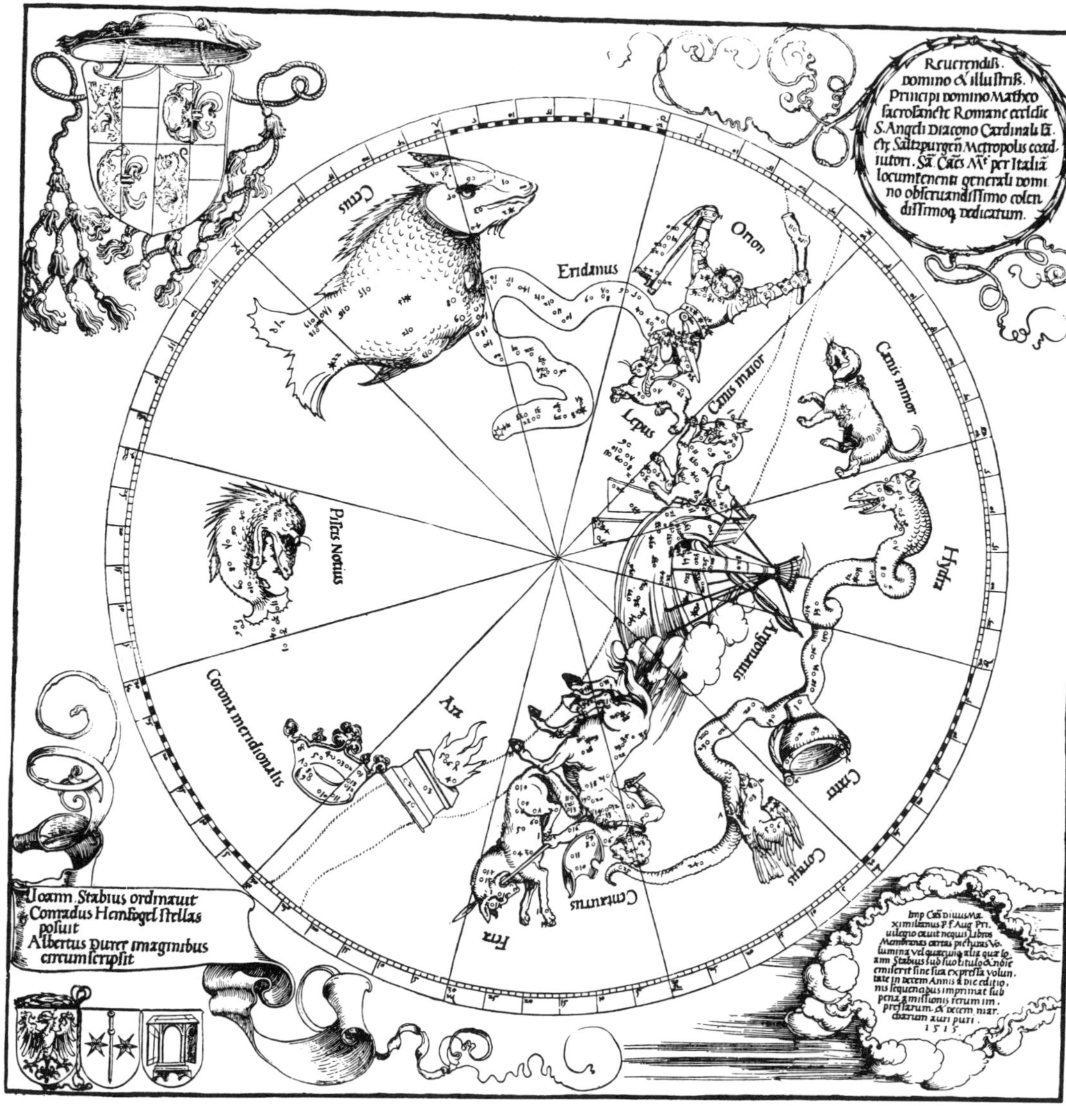

Star map of the Southern Hemisphere, also engraved by Albrecht Durer in the early 1500s.

Star maps use a system of coordinates to locate a particular star or planet. In astronomical mapping, the coordinate that corresponds to longitude is called **right ascension** (azimuth). The coordinate that corresponds to latitude is called **declination** (altitude). By determining the right ascension and declination of a star, its position can be precisely fixed on a star map.

Right ascension is measured west of the point in the sky where the sun is on March 21, or the first day of spring, when the sun crosses the equator. Declination is the angular distance of an object from the **celestial equator**. On most star maps the right ascensions are marked around the edge of the map. The complete circuit around the sky is divided into 24 hours, and these are divided further into minutes and seconds.

Imagine that you are standing at the center of the Earth and that you are able to see both the lines of latitude and longitude projected onto the Earth's surface. Now imagine that the dome of the Earth's surface is the sky. The lines of latitude and longitude would then become lines of declination and right ascension.

The pattern for the star map included on page 64 is marked according to hours and minutes. It is also marked with degrees to correspond with the design of your rotating quadrant. Declination, or altitude, is marked in degrees along the sides of the segments radiating out from the center of the star chart.

RIGHT ASCENSION

Corresponds to the longitude on Earth and measures the location of a star in terms of hour circles which pass through the north and south celestial poles. The celestial equator is divided into 24 hours or equal parts.

DECLINATION

Corresponds to the latitude on Earth. Declination tells how far north or south of the celestial equator a star is located.

CELESTIAL EQUATOR

A great circle on the celestial sphere that is directly above the Earth's equator.

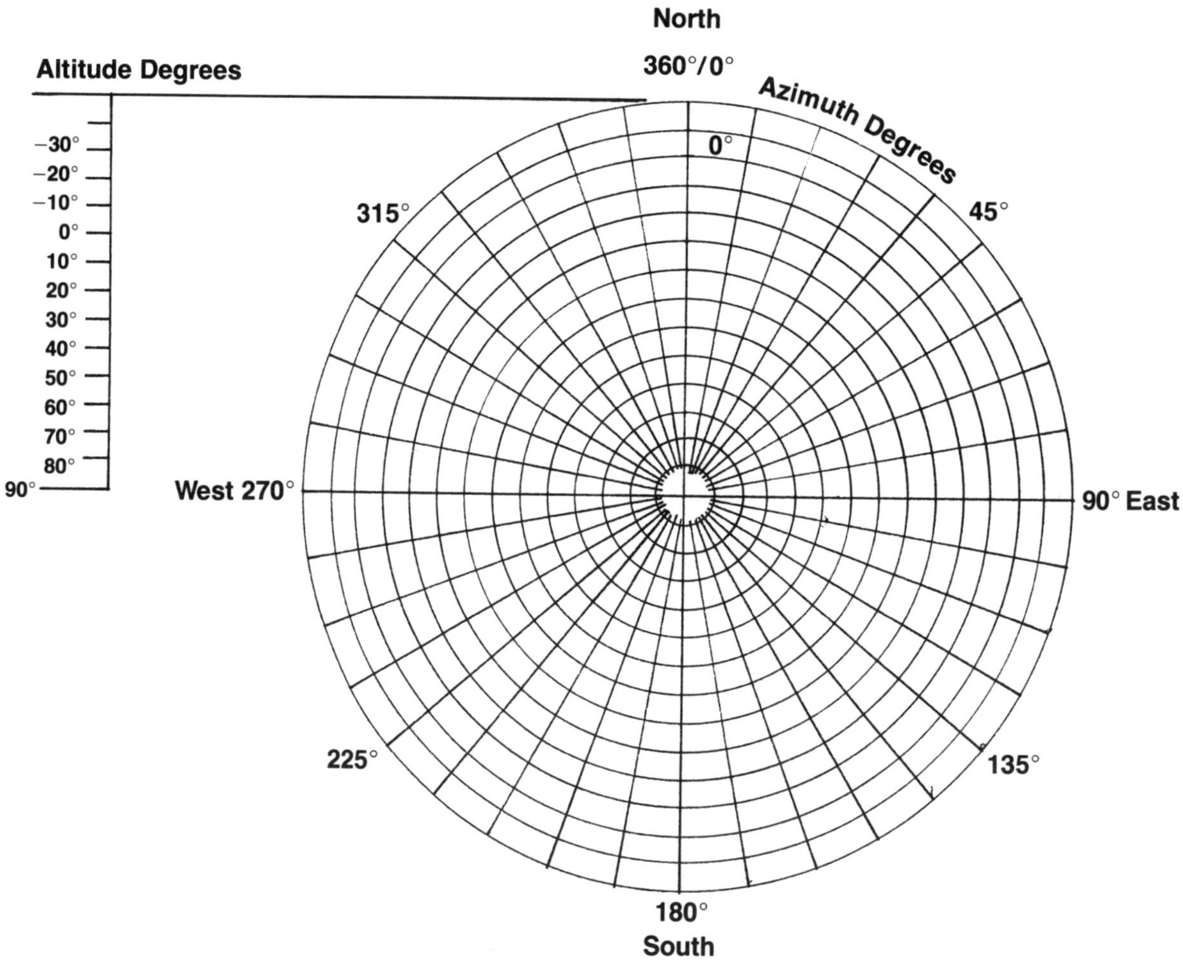

To begin recording your observations, copy the pattern for the star map on page 64. Make 4 copies of the pattern and assemble them to form a circle like the star map above. Then sight a star. Read the right ascension, or azimuth measurement, from the base of the quadrant. Locate this measurement on the star chart grid. Then read the declination, or altitude measurement, of the star from the protractor and locate it along the marked line radiating from the center of the star map. For example, a star located at 100 degrees right ascension and 40 degrees declination would be placed on a star chart as shown below.

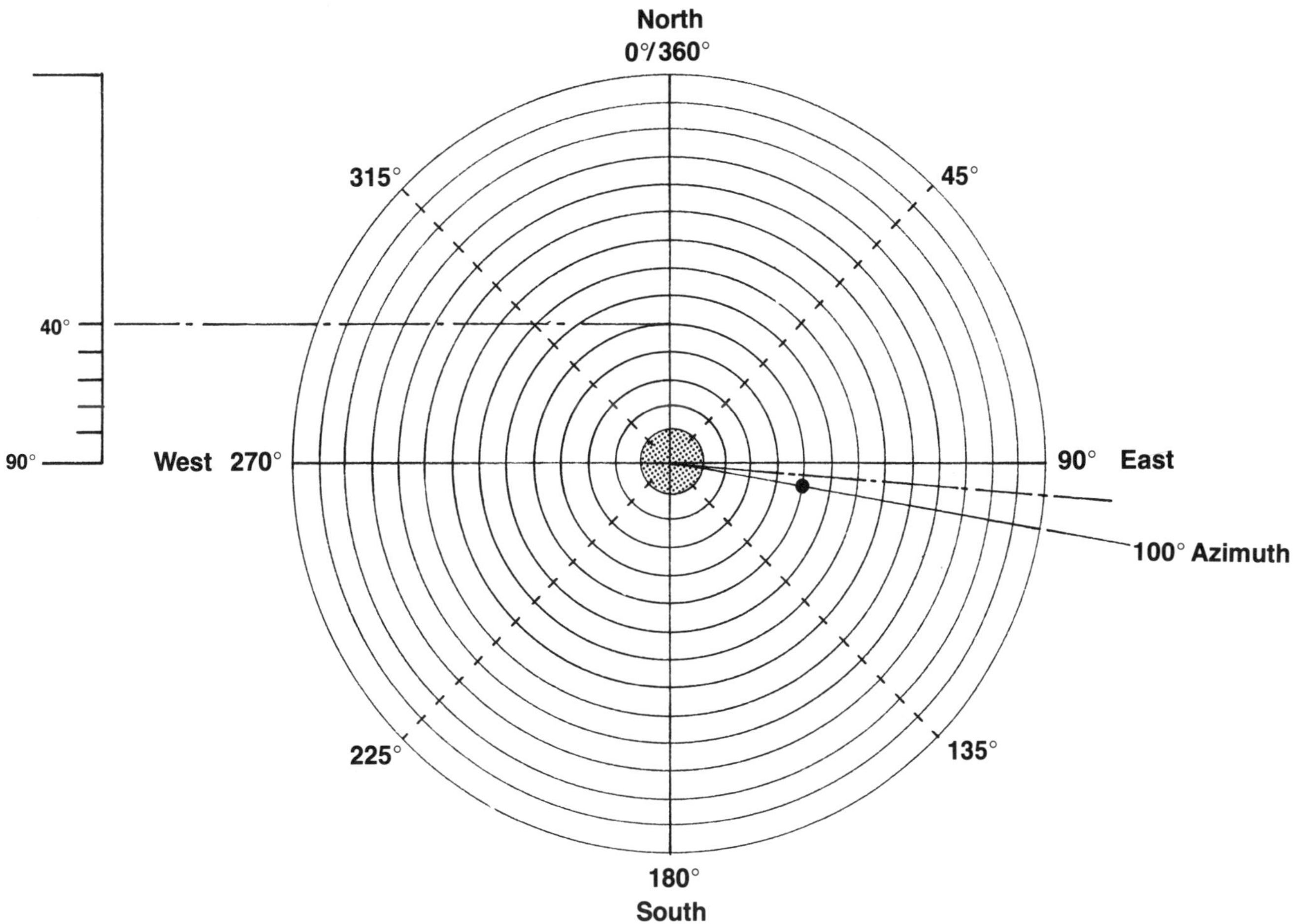

The star map or chart you have made is meant to be used to record as many stars and planets as you can observe on one night. If you live in a city, you will not be able to see the dim stars because the city lights block them out.

Because the Earth moves around the sun, the positions of the stars you observe on one particular night will shift slightly each night. The reason for this is that the Earth takes 23 hours and 56 minutes to make one complete rotation. Since this is four minutes less than the 24 hours in a day, any particular star you are observing will rise four minutes earlier each night.

Because the movement of the Earth causes these shifts in the location of the stars, they will remain at about the same distance from one another. All the stars will seem to rotate around the North Star though, which does not appear to move because it lies directly above the North Pole.

Since the seasons change as the Earth moves, the stars and constellations you can observe on a winter night will not be the same ones that you can see on a summer night. A different part of the sky will be seen each night, as the Earth completes its yearly orbit around the sun.

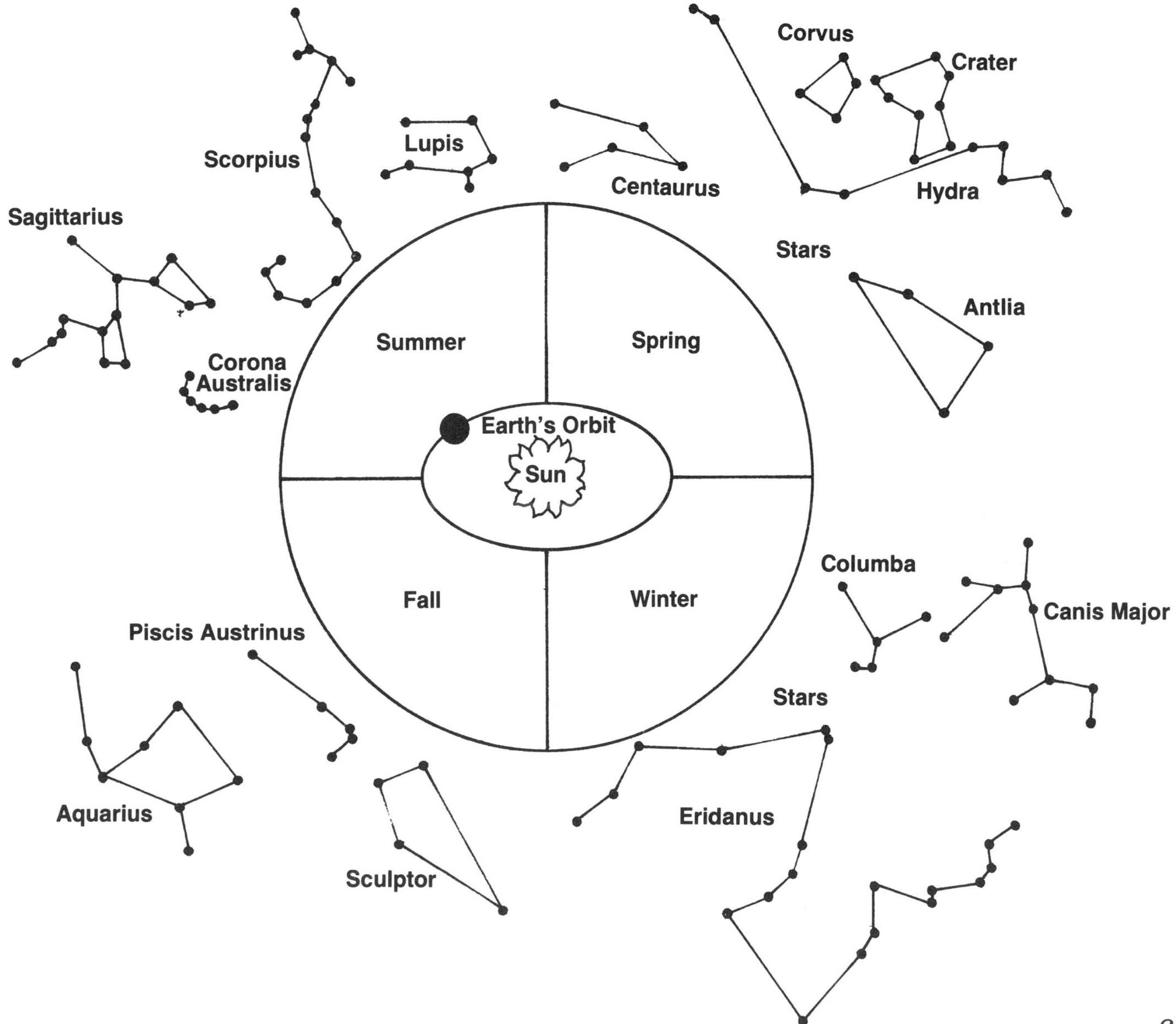
Corvus
Crater
Scorpius
Lupis
Centaurus
Hydra
Sagittarius
Stars
Antlia
Summer
Spring
Corona
Australis
Earth's Orbit
Sun
Fall
Winter
Columba
Canis Major
Piscis Austrinus
Stars
Aquarius
Eridanus
Sculptor

STAR MAP PATTERN
(Make four copies.)

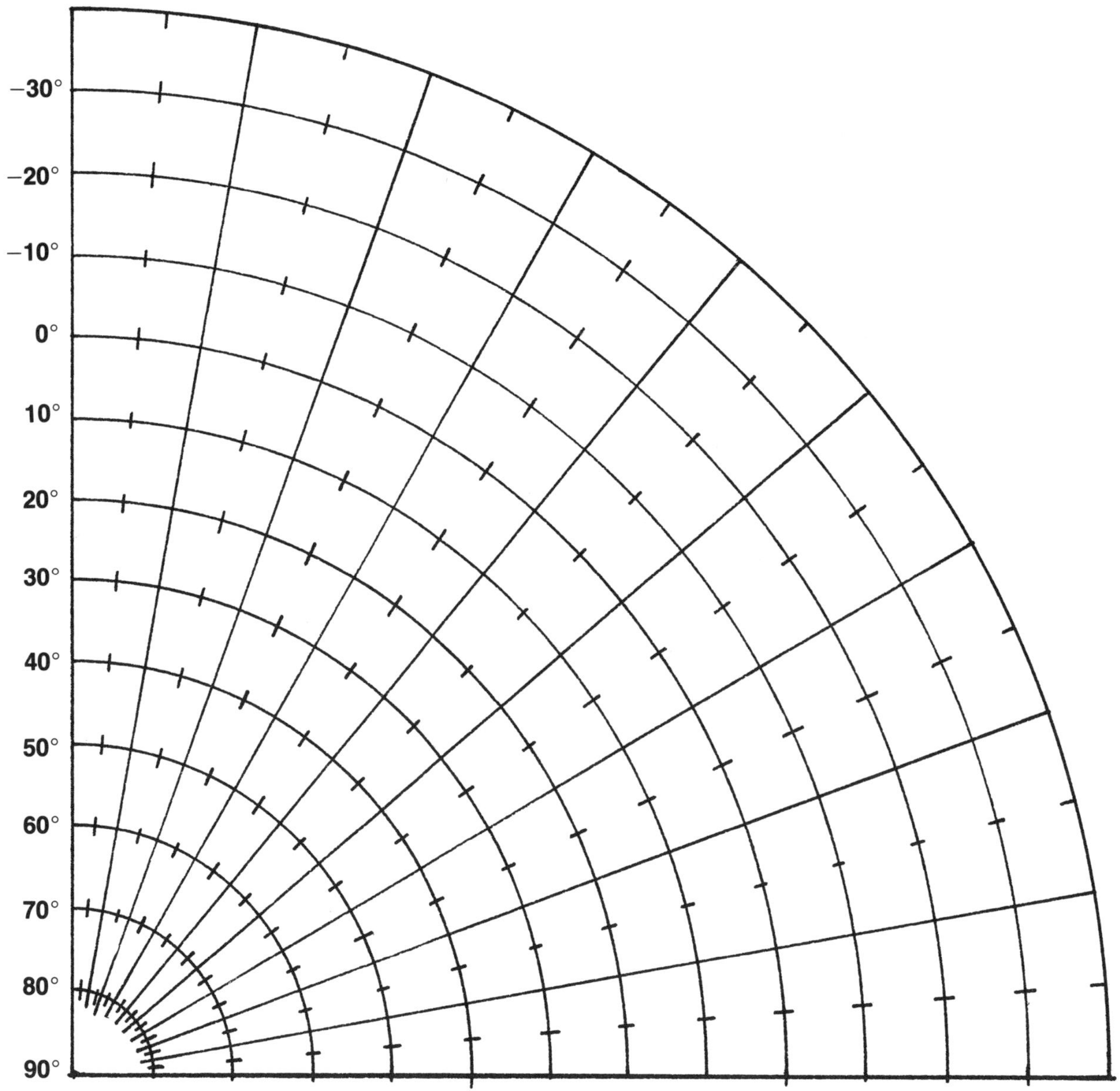

Each 15° = 1 Hour

With the fall of the Roman Empire during the A.D 400s, the cosmology of Aristotle, and the writings and star tables of the astronomer Ptolemy, were lost and forgotten. Not until the 650s, when the Arabs conquered Egypt, were they rediscovered. By the 800s, Arabic works on astronomy were appearing. Arabic astronomers were using quadrants and astrolabes to make measurements and computations, and to help them draw up complete star tables.

It was not until the 1200s, however, that Arabic translations of Aristotle made their way through Spain into France. In Spain, King Alfonso X gathered together astronomers to construct a new set of star tables. Completed in 1252, the "Alfonsine Tables," were to be used throughout Europe for at least 300 years.

After the works of Aristotle were translated into Latin they began to be taught in both universities and church schools throughout Europe during the Middle Ages. Since the Aristotelian and Ptolemaic notions—that the heavenly bodies were ruled by God—fit in with what many religious leaders believed at that time, it was felt that to question Aristotle was to actually question the existence of God. For these churchmen, God was the force that ruled and moved the universe. For them, the universe was a sphere filled with all sorts of heavenly matter, including all types of beings such as seraphims, thrones, archangels and angels. The heavens were not an empty place.

One of the most interesting theories suggested during the 1400s was made by Nicholas Cusa, a German Cardinal. Cusa tried to combine the theories of church and science. He concluded that the Earth could not be the center of the universe, since the center of the universe would have to be occupied by the most important part of the entire system—God.

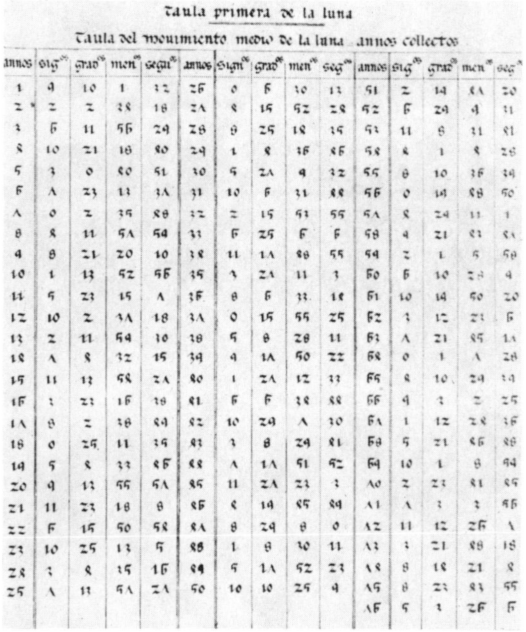

Taula primera de la luna

Taula del mouimiento medio de la luna annos collectos

A page from an 1862 reprint of the *Alfonsine Tables* of King Alfonso X of Spain. *Courtesy of the Rare Book Collection, Special Collections, Richter Library, University of Miami*

An illustration of an early astonomer. From Jost Amman and Hans Sachs, *Eygentliche Beschriebung Aller Stands Auff Erden . . .* (1568).

Unfortunately, many of the medieval scholars were merely quoting the theories of Aristotle and following the star tables of Ptolemy and the Alfonsine Tables without questioning them. For over 1,000 years, almost nothing new was learned in astronomy.

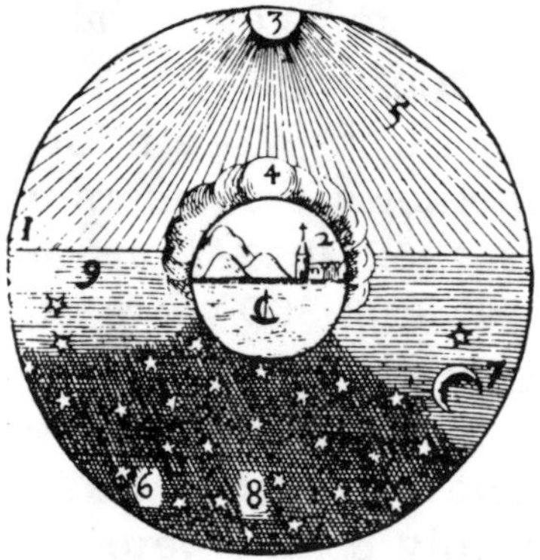

An illustration of the geocentric universe published in Comenius' *Orbis Sensualium Pictus* in 1659. The heavens (1) are wheeled about and encompass the Earth (2). The sun (3), the clouds (4), and the rays of sunlight (5) make up the day. On the other side is darkness (6) with the moon (7) and the stars (8). Between day and night is the twilight (9).

It was not until the mid-1500s, when Nicholas Copernicus developed his theory that the Earth is a planet that moves around the sun, that modern-day astronomy was born. When Copernicus was born in Torun, Poland, in 1473, most people still believed, as they had for over 1,000 years, that the sun rotated around the Earth every day. Today we are able to actually look back at the Earth, spinning in its orbit around the sun, from the moon or satellites. But in the late 1400s, people still believed in the Ptolemaic system. After all, if the Earth moved, why didn't they feel it?

Copernicus' new system of cosmology was outlined in his book, *De Revolutionibus Orbium Coelestium* (On the Revolutions of the Heavenly Spheres), which was published in 1543. His **heliocentric** theory was based upon his belief that the Earth is a heavenly body, a planet of the sun, and that it is constantly moving. Copernicus believed that people on Earth do not see the Earth moving because they are moving with it. His theory of the universe used the Earth's movement through space to explain the movement of the sun and the stars.

HELIOCENTRIC

A system of cosmology which places the sun in the center of the universe.

When Copernicus was a young man in Torun, it was a busy trade center. Tradesmen, clerics and scholars traveled all over Europe and brought back news of new ideas to Torun. Copernicus himself traveled widely. At 18 he began his undergraduate studies at the university in Cracow, Poland. While Copernicus was there, Columbus completed two voyages to the New World, discovering both Cuba and Jamaica. It was at Cracow that Copernicus learned how to use observational instruments and how to make astronomical computations.

In 1496, Copernicus traveled to Italy to study law, and then medicine. He continued his study of mathematics and astronomy at the University of Bologna, under the famous astronomer Novara. In Italy, Copernicus himself lectured on mathematics and made his first accurate observations of the stars.

Copernicus returned to Poland in 1503. He became a canon at Frombork Cathedral. He was the administrator of the cathedral estates, and a doctor, but not yet a well-known astronomer. Copernicus had plenty of time to write and observe the stars from a corner tower of the castle, but there were few people in Frombork that he could talk to about his theories of a sun-centered system of cosmology.

By 1515, Copernicus had developed the basic theories of his heliocentric system and had prepared handwritten discussions of them, which he sent to friends and astronomers throughout Europe.

Nicholas Copernicus (1473-1543) *Courtesy Dibner Library, Smithsonian Institution*

Even though his reputation was growing as an important astronomer, he preferred to stay in Frombork and study quietly and slowly. Because he made most of his own instruments, he could only make rough measurements and had to rely on the published calculations of earlier astronomers.

Perhaps Copernicus was unsure of some of his observations. Or perhaps he was afraid that his sun-centered system would cause a great deal of trouble in a world where many churchmen, scholars and even scientists still believed that the sun whirled around the Earth. Whatever his reasons, Copernicus was very hesitant about publishing his theories in a book which could be widely circulated. There was no printer in Frombork, so when a young mathematician from Germany, Georg Joachim, came to learn about Copernicus' theory, and convinced him to publish his book, it had to be taken to the city of Nuremburg, Germany, to be printed.

Copernicus saw a copy of this work in May 1543, just before he died. Even though most of it had been written 15 years earlier, other astronomers and scholars only had a chance to see it after his death. Copernicus' theory began a revolution in astronomy and laid the foundations for many of our modern scientific theories.

For example, before Copernicus, people answered the question "What is night?" in this way: the sun rises in the east, passes overhead, descends in the west and goes behind us.

But in Copernicus' system the sun does not move. Night occurs because the Earth rotates every 24 hours. When the light which is always coming from the sun is blocked by the Earth as it moves in its orbit it is night on part of the planet.

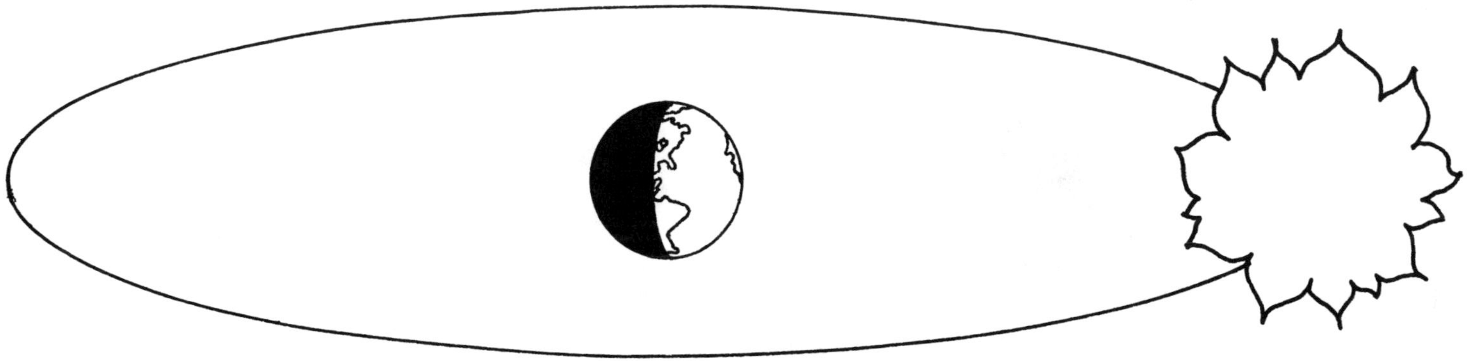

The sun rising in the east.

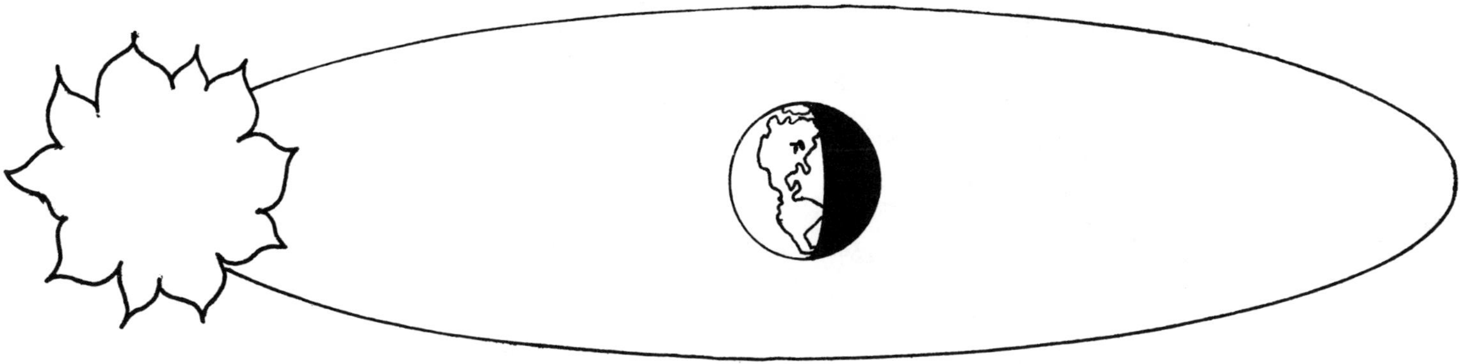

The sun has passed overhead and is descending in the west to create night on one side of the Earth.

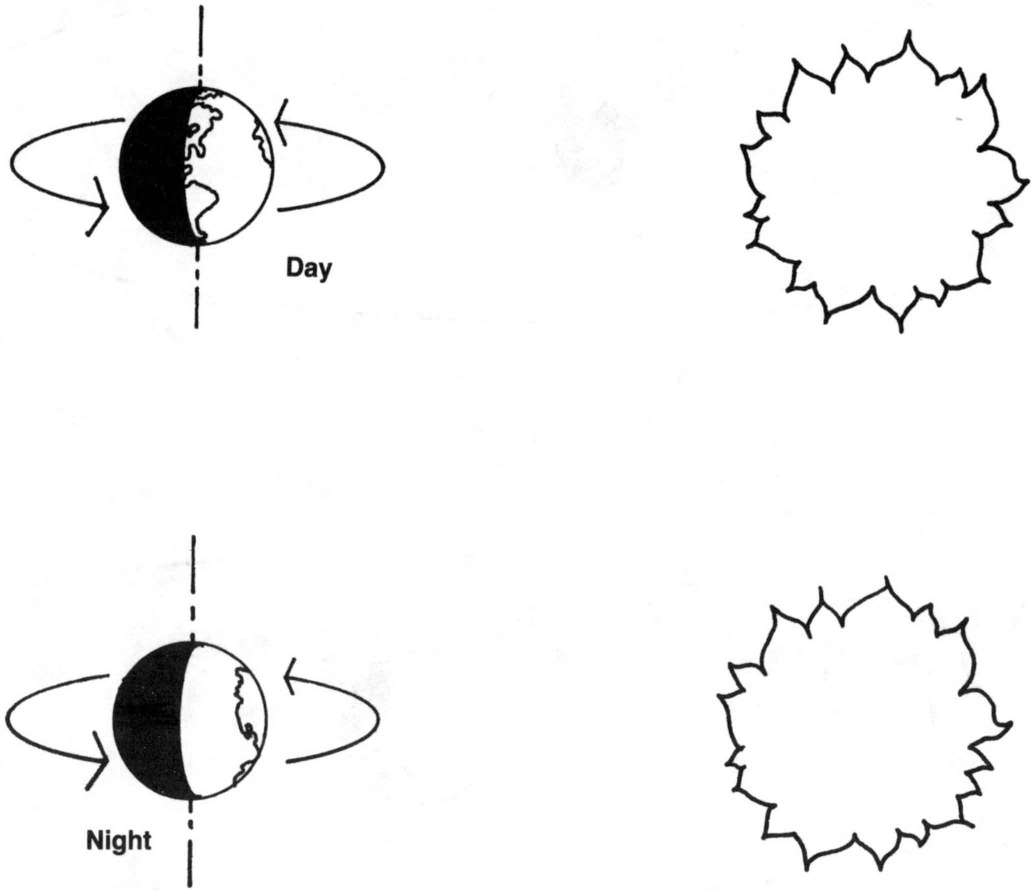

In Copernicus' system the Earth rotates on its own axis, which causes it to be day, and then night, on either side of the planet.

A simple demonstration of how night follows day was included in many science and geography books during the 1800s. You can repeat this demonstration by setting a lamp, which represents the sun, on a table. An apple or any other small, round, soft object can be used to represent the Earth. Push a knitting needle or stick through the apple to form the Earth's axis. Set the apple at the edge of the table. When it is rotated on its axis, the rays from the lamp will strike the apple, as the sun's rays strike the Earth, creating day and night.

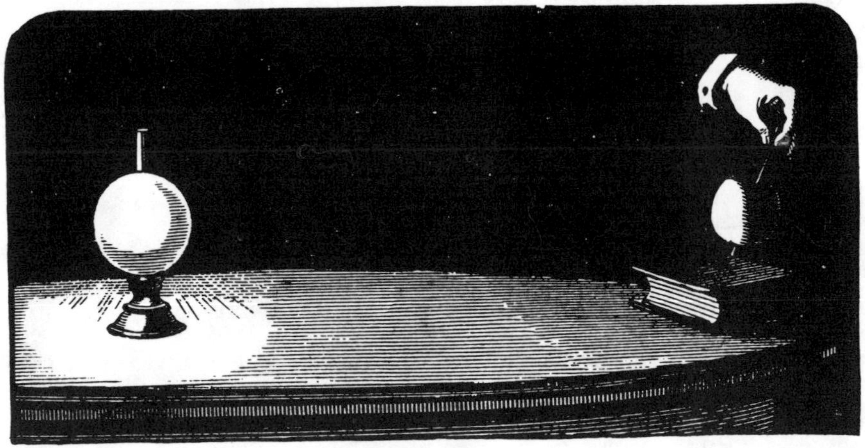

The "succession of day and night" demonstration, from James Monteith's *Comprehensive Geography*, (1872).

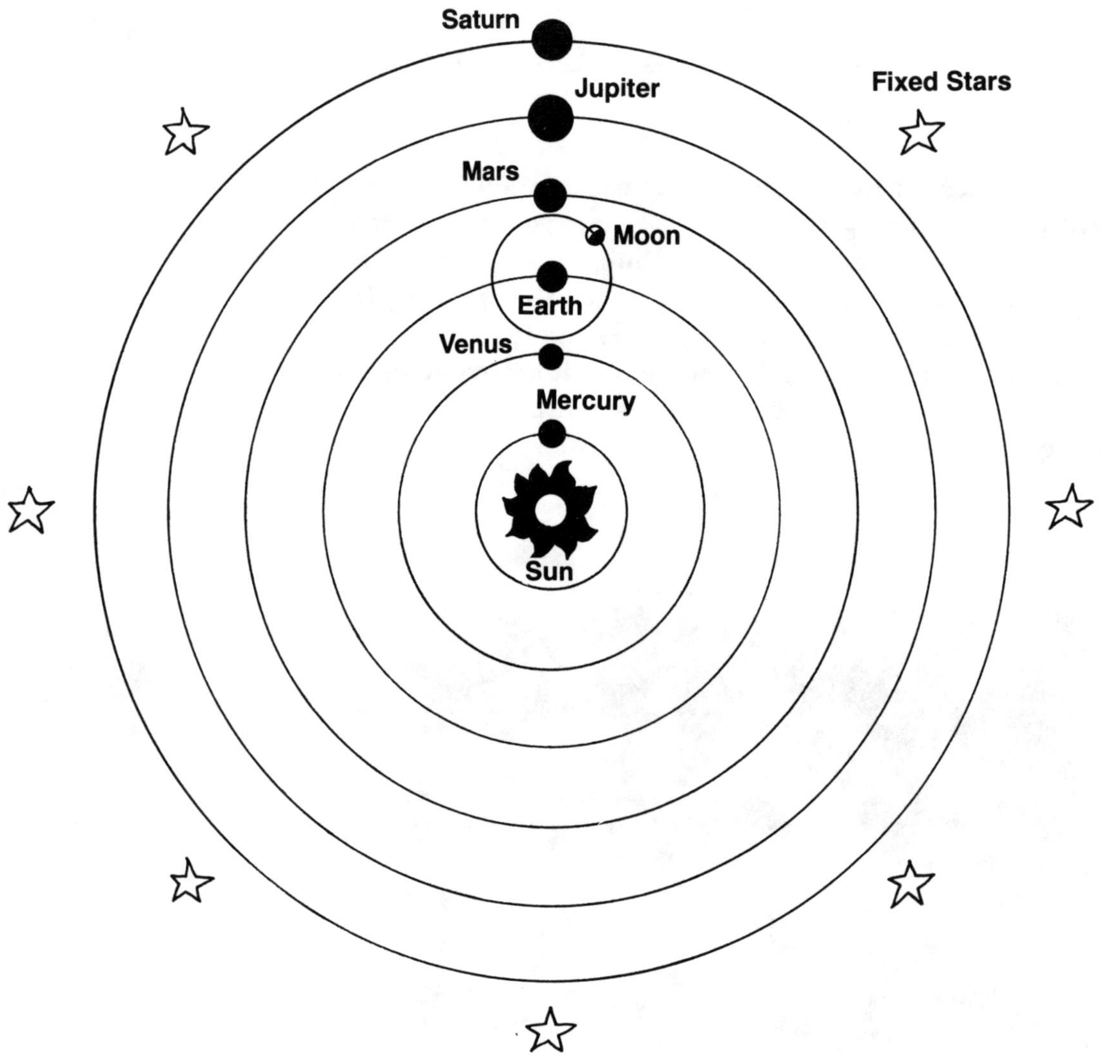

The solar system according to Copernicus. The Earth and other planets rotate around the sun in their circular orbits.

The main points of Copernicus' cosmology, which were described in his 1515 commentary and later published in his book in 1543, challenged astronomers for at least 100 years after his death. Copernicus' main points were:

1. The Earth is not the center of the universe.
2. The center of the universe is the sun.
3. The Earth rotates around the sun, one complete rotation each year. The sun does not actually move, but only appears to as a reflection of the Earth's movement.
4. Day and night occur because the Earth rotates on its own axis, one complete rotation each day. For this reason, the stars we can see from the Earth appear to travel from east to west.
5. All the other planets also rotate around the sun, which is why they appear to move around the Earth. The sun itself appears to change its position, because throughout the year the Earth is rotating around it.

Copernicus' work as an astronomer was mostly theoretical. His theories were not based so much on his own experience or observations, but on his belief in the simplicity and harmony of the universe. During his entire lifetime, he made probably only 70 astronomical observations. For this reason, much of his work seems to have been based on observations made by earlier astronomers. Therefore, even though he developed a heliocentric system, it was not necessarily accepted as the truth by other astronomers. In fact, Copernicus' system did not even fit the laws of physics that were known at that time. For this reason, Tycho Brahe, the first of the great modern systematic astronomers who lived and worked after Copernicus' death, still believed in a geocentric system of cosmology.

Tycho Brahe (1546-1601) *Courtesy Dibner Library, Smithsonian Institution*

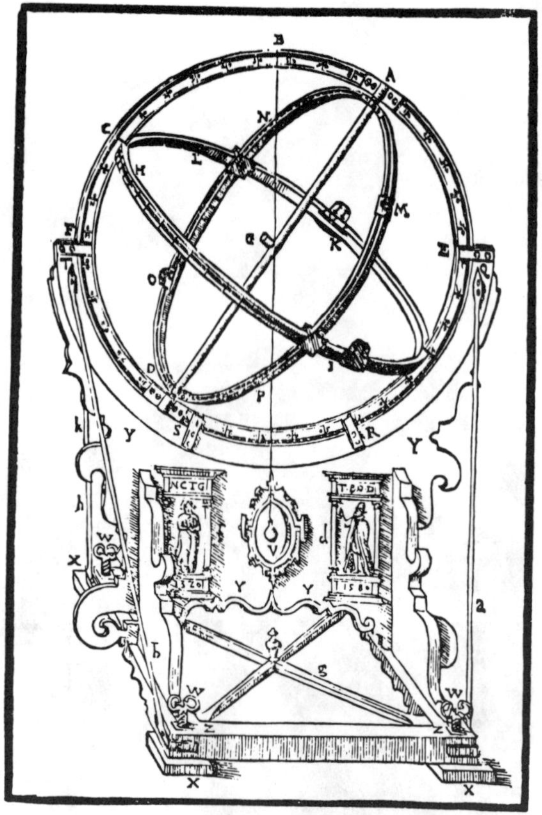

An equitorial armillary sphere, from the late 1500s, used by the astronomer, Tycho Brahe. From his *Astrononiae Instauratae Mechanica*

Tycho Brahe was born in 1546, three years after the death of Copernicus. He was the son of a wealthy Danish nobleman. Brahe began to study astronomy when he was fourteen years old. He was also very interested in mathematics. In fact, as a student he fought a duel with one of his classmates over who was the best mathematician. Unfortunately, part of Brahe's nose was sliced off during the contest. Therefore, he always wore a false nosepiece, made of gold and silver, over the missing part.

Brahe went to universities in Germany until he was 26 years old. At that time he began to make the series of thousands of careful, systematic observations of the stars that would keep him occupied until the end of his life. In November 1572, Brahe discovered a new star, in the constellation of Cassiopeia, that was even brighter than Venus. Everyone wondered how it was possible for a new star to appear in a universe that was supposed to be fixed and unchanging! Brahe made exact observations of the star and published them the next year in his book called *De Nova Stella* (The New Star). Even though he could not prove how the new star came into being, his observations of it proved that the universe was not unchanging. Today we know that Brahe had actually observed a supernova, or the explosion of a star.

In 1576, when he was thirty and already famous, Brahe returned to Denmark where he set up an astronomical observatory on the island of Hven, which is located between the coasts of Sweden and Denmark. The Danish king, Frederick II, had given him the island and paid for the three-story castle observatory. It was the finest that had ever been built in Europe. Brahe called it "Uraniborg," after Urania, the Greek god of astronomy. A chemical laboratory, paper mill, printing press, and museum library were all included as part of the castle. Observational platforms were located at various points throughout the castle, and sections of the roof could be moved so that he could observe the heavens with his instruments.

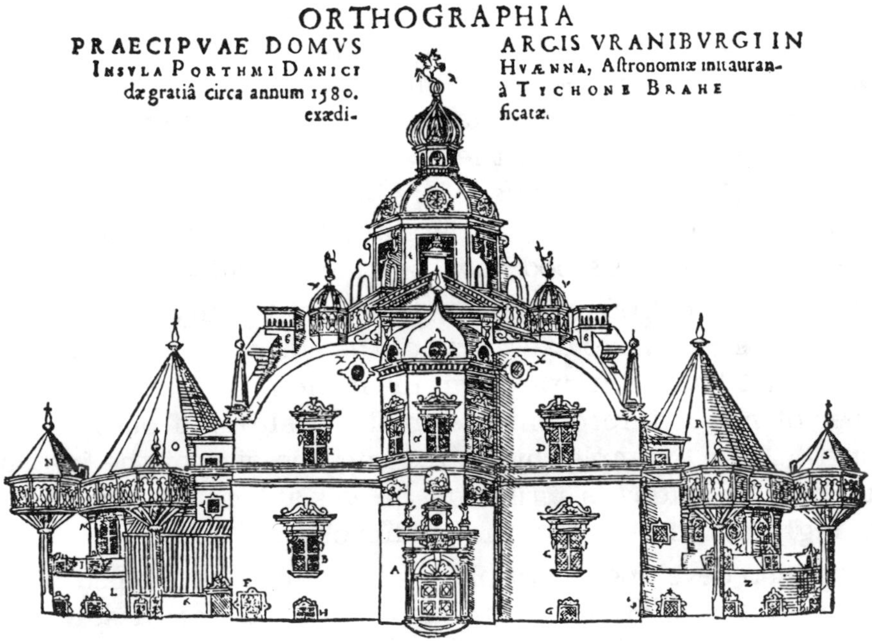

An illustration of Uraniborg, Tycho Brahe's castle observatory, published in a collection of his letters.

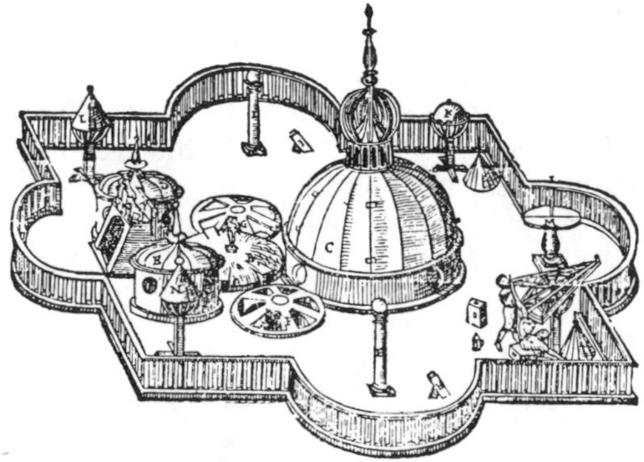

Stjerneborg, Brahe's star castle, which was built below ground level. Tycho Brahe, *Astrononiae Instauratea Mechanica*

About 400 feet away from the main castle observatory, Brahe built a second observatory called *Stjerneborg,* or the "star castle." It was below the ground to keep the wind from disturbing the instruments with which Brahe made his observations. It is probably not surprising that the peasants who lived on the island thought that Brahe, with his magical castles and instruments, was a sorcerer or magician.

This illustration from the 1500s is actually a wall painting at Uraniborg. It shows Tycho Brahe observing the stars with the aid of his two huge wall quadrants.

Brahe's astronomical observations were completed entirely without the aid of telescopes, which were invented shortly after his death. Instead, he used improved versions of such ancient astronomical instruments as quadrants, armillary spheres and paralactic rulers. All of these instruments used simple metal sights. Basically, Brahe was trying to determine the exact position of as many stars as possible, and also the motion of the planets that could be seen with the unaided eye (Mercury, Venus, Mars, Jupiter and Saturn). It was his belief that men could never understand the true purpose of the universe until they understood how the planets and the stars worked.

Brahe's main contribution to modern astronomy was that he compiled the most accurate catalogue of the stars, and their positions in the sky, that had been drawn up until then. In addition, he made a number of major improvements in the designs of observational instruments. For example, by increasing the size of the instruments with which he made his sightings, he was able to improve the quality of the sighting and the accuracy of his observations.

Tycho also developed portable instruments that helped him in his astronomical observations. The most important of these was the astronomical sextant. The sextant really did the same job as the quadrant, but it was much more practical. With a fixed and movable sight, it could be used, by two observers, to determine the angular distance between any two objects in the sky.

As astronomical instruments such as Tycho Brahe's became more sophisticated and accurate, the study of the stars was put to new and practical uses. Astronomical observations were used to help draw up maps of different parts of the Earth, and to help sailors navigate across the seas. Systematic observation of the sun and stars was being used to define not only the shape and size of the Earth more accurately, but also to understand the distance between different points on the Earth's surface.

Actually, the use of astronomy to understand the nature of the Earth began with the Greeks. In about 350 B.C., Aristotle argued that the Earth must be round, since the outline of its shadow on the moon during eclipses was round. The size of the Earth, and the fact that it is round, was proven by the Greek mathematician and astronomer, Eratosthenes of Cyrene, in about 250 B.C..

Eratosthenes did this by measuring the angle of altitude, or the angle of the sun above the horizon, at noon on the day of the summer solstice in Alexandria, Egypt. Since he was from Cyrene (latter-day Aswan), he already knew that the sun was directly overhead at noon on that day. When a gnomon was stuck in the ground in Aswan, no shadow was cast since the sun was directly above. When Eratosthenes compared the altitude of the sun at Alexandria with that at Aswan, he was able to determine that Aswan was about seven degrees farther north than Alexandria. He was actually measuring the difference in latitude between the two towns.

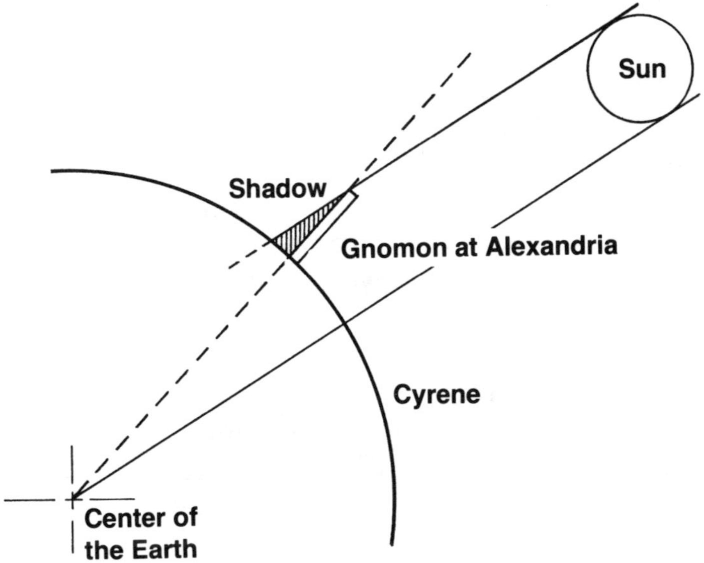

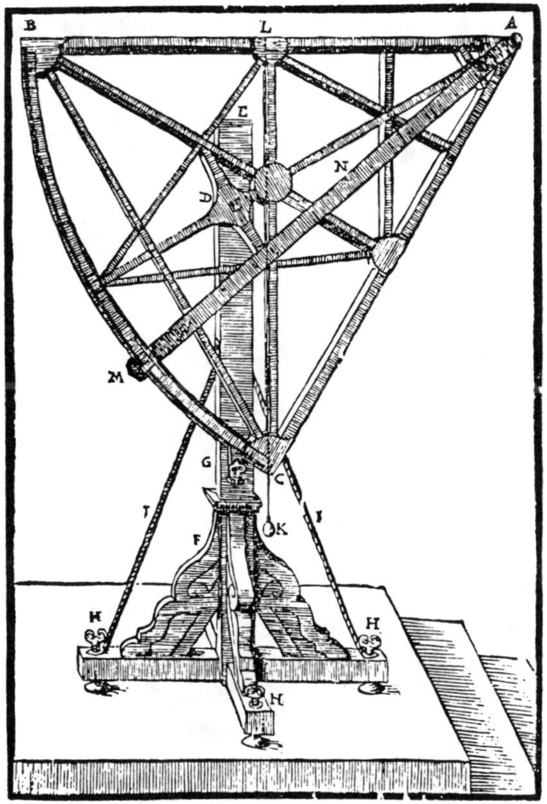

Tycho Brahe's astronomical sextant, which had an extra pair of sights and was portable, even though it had a radius of five feet. Tycho Brahe, *Astrononiae Instauratea Mechanica*

Eratosthenes also knew that the two towns were about 500 miles apart. He divided the 500-mile distance between them by the seven degree difference in latitude between them. The answer of about 70 miles was the approximate distance for one degree of the **circumference** of the Earth. Since Eratosthenes concluded that the Earth is curved because of the differences in the sun's elevation in the two towns, he was able to multiply the number 70 by 360, or the number of degrees in a circle. His answer of about 25,715 miles was his measure of the circumference of the Earth. Today we know the circumference of the Earth at the equator is about 25,000 miles, and each degree of latitude is about 69 miles.

CIRCUMFERENCE

The distance around a circle or sphere. Every point on the circumference of a circle is the same distance from the center of the circle.

By using his astronomical instruments and mathematics, Eratosthenes had been able to determine that the Earth was round and to obtain a figure for its circumference. Unfortunately, this figure, and even the knowledge that the Earth was round, was forgotten by many people during the early Middle Ages, or from about A.D 400 to 1,000.

Throughout the Middle Ages, most sailors relied on the stars and sun to determine their position and course. Therefore, they could only sail from May through September, when the stars were most likely to be visible. Sailors did have some navigational tools and instruments to help them travel. The portulan chart, which was a very accurate map of coastlines, came into use during the 1200s. The **knotted line** and the **traverse board** were also used to determine a ship's speed and to plot its course.

KNOTTED LINE

A marked cord, with a log tied on its end, that was thrown overboard the ship. The cord was paid out for a specific amount of time. By counting the number of knots paid out in a set period of time, sailors were able to determine the speed of their ship.

TRAVERSE BOARD

A board with holes set in concentric circles. The center hole represented the ship's position and pegs were placed in the other holes to plot the ship's course.

Most of the maps of the world that were drawn up during the Middle Ages look strange to us today. Some show all the land in the middle, surrounded by water, while others show the Earth having four corners. It was only in 1409, when Ptolemy's works were translated into Latin, that his map of the world, which was based on astronomical observations, was rediscovered. Ptolemy's map showed a spherical Earth with its surface divided into degrees of longitude and latitude.

An interest in combining the sciences of geometry, mathematics and astronomy was reviving in Europe at this time. Both sailors and map makers began thinking of the Earth in terms of latitude and longitude, instead of coastlines and landmarks.

Sailors were also interested in making astronomical instruments easier to use at sea. Among the instruments they used to make observations were the astrolabe and the quadrant. Even though both were modified to be used at sea, it was not until the 1700s that a really practical method of finding longitude at sea was discovered. Until that time there was no instrument capable of measuring the altitude of the stars with any degree of accuracy. The invention of the reflecting quadrant, or sextant, in the early 1700s made the required accuracy possible.

The new instrument combined the telescope with reflecting mirrors and made it possible for navigators to measure the altitude of a star. Thus, the position of a ship could be determined accurately to within 100 yards.

The rotating mirror, attached to the movable arm of the sextant, reflected the light from a heavenly body onto the horizon glass. The top half of the horizon glass was clear and the lower half was a mirror. The navigator could then read the altitude of the heavenly body off of the degree scale at the bottom of the instrument. On the sextant, the degree scale represents the arc of 1/6 of a circle, or 60 degrees. The size of the degree scale may vary in different sextants.

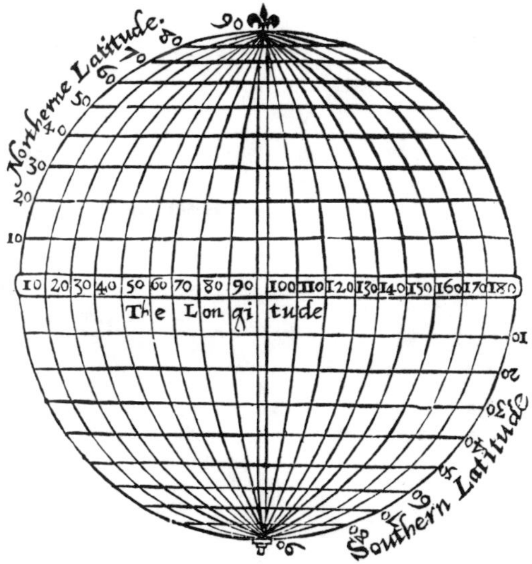

An illustration of latitude and longitude from a book on navigation published in 1594.

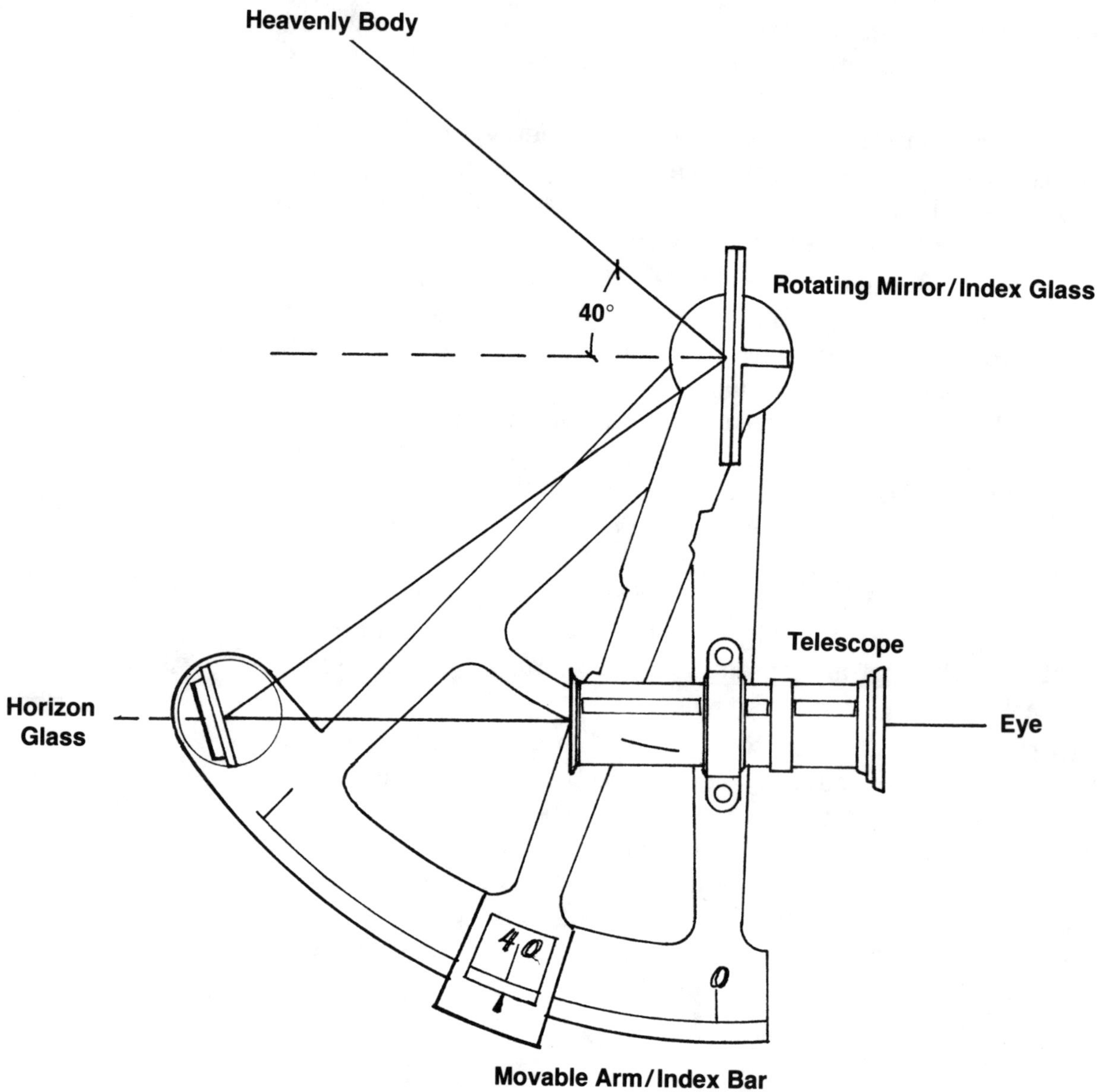
Heavenly Body
40°
Rotating Mirror/Index Glass
Telescope
Horizon
Glass
Eye
40
0
Movable Arm/Index Bar

Ironically, the machinery necessary to accurately mark the degree scales on small portable sextants did not exist when they were invented. Marks made by hand were not accurate enough. Not until the mid-1700s was it realized that the machinery used to mark the divisions on a clock wheel could be adapted to mark sextants.

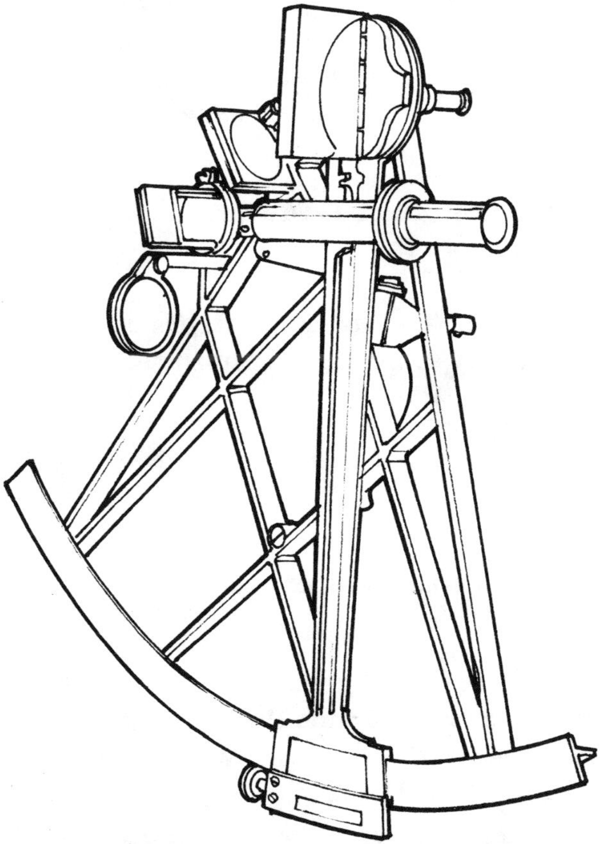

Sextant made in London in about 1790 by John Dolland, a famous optician and instrument maker. The instrument measures about one foot from the base to the top. The telescope is about five inches long.

You can make a working sextant that uses a sighting device in place of a telescope and has a degree scale of 90 degrees. Even though your sextant will not be as accurate as one whose degree scale is machine-made, it will still give you fairly accurate measurements of the altitude of stars and other objects. You will need the following materials:

Two 8″ x 11″ pieces of 1/12″-thick cardboard
1 soda straw, 3″ long
1 small mirror, 1″ x 2″
1 small mirror, 1″ x 1″ (Small pocket mirrors may be cut with a glass cutter, or you may ask someone in a hardware store to cut the two pieces of mirror for you.)
One ¼•20 bolt and nut
White glue
X-Acto knife
Scissors
Paper punch to make small circular holes

Copy the sextant patterns on pages 88-90 onto tracing paper. The patterns may also be photocopied. Mount the patterns on the pieces of cardboard and cut out the pieces of the sextant with the X-Acto knife.

Glue the sighting support onto the main body piece of the sextant. Then glue the soda straw along the edge of the sighting support as shown on page 83. Next glue the eyepiece to the end of the soda straw, and the mirror mount to the body of the sextant.

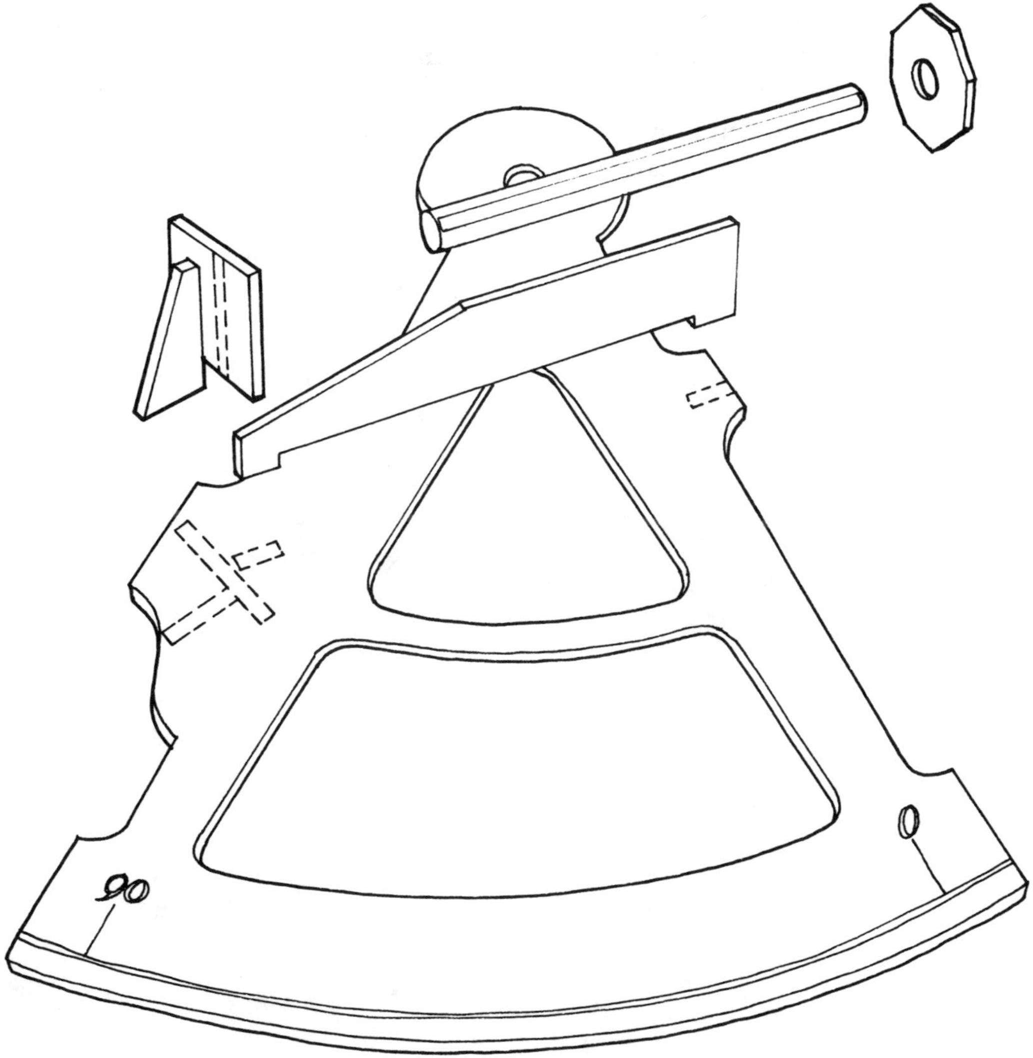
90
0

Cut out the sextant's arm with the X-Acto knife. Glue the guide pieces for the arm onto the bottom of the arm piece. Slide the arm piece onto the body of the sextant so that the top of the sextant's arm piece runs along the base of the sextant's body. Insert the bolt through the hole in the top of the sextant's arm and through the hole in the top of the body. Set the nut in the bolt and tighten it. Glue the mirror mount onto the top of the sextant's arm, partially covering the head of the bolt.

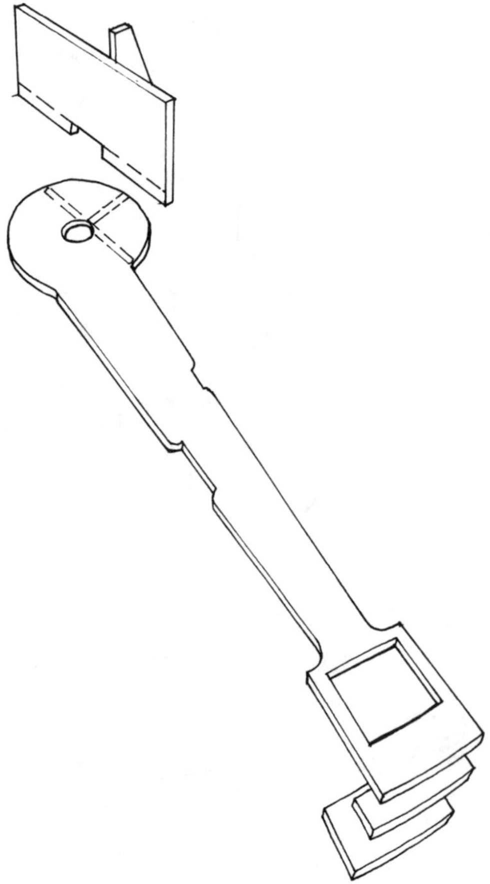

Draw a line with the X-Acto knife or a pencil on the back of the 1″ x 1″ mirror, dividing it in half. Scrape the backing off half of the mirror with the X-Acto knife. Glue the mirror onto the mirror mount so that it covers half of the mount, leaving the clear glass part of the mirror overlapping the edge of the mount.

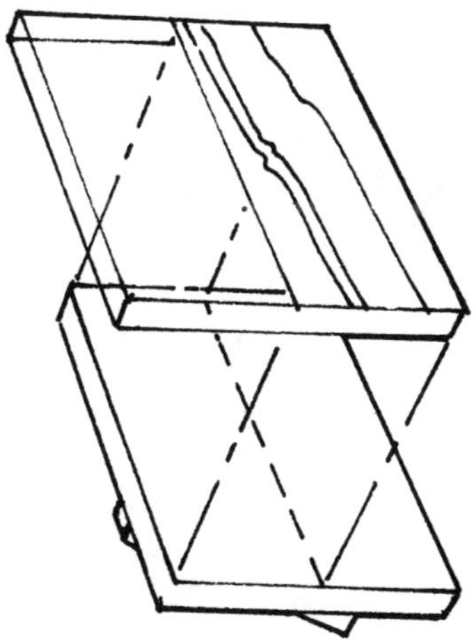

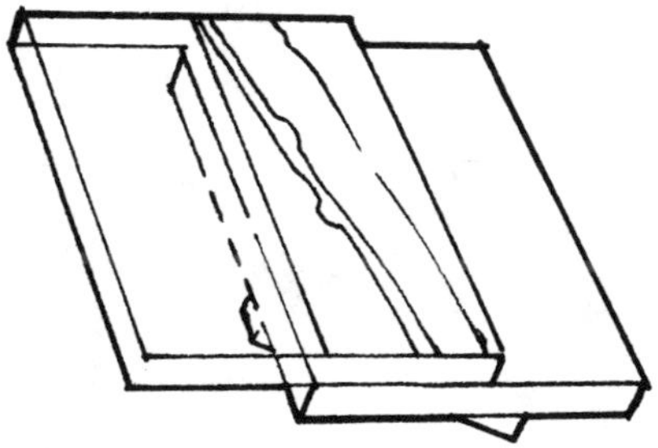

If you look through the sighting hole of the sextant you should be able to see the mirror in the right half of the sighting piece and the clear half of the mirror in the left side.

Glue the 1″ x 2″ mirror to the mounting piece at the top of the sextant's arm.

Glue the 3 support pieces onto the back of the sextant's body as illustrated. Your sextant is now complete.

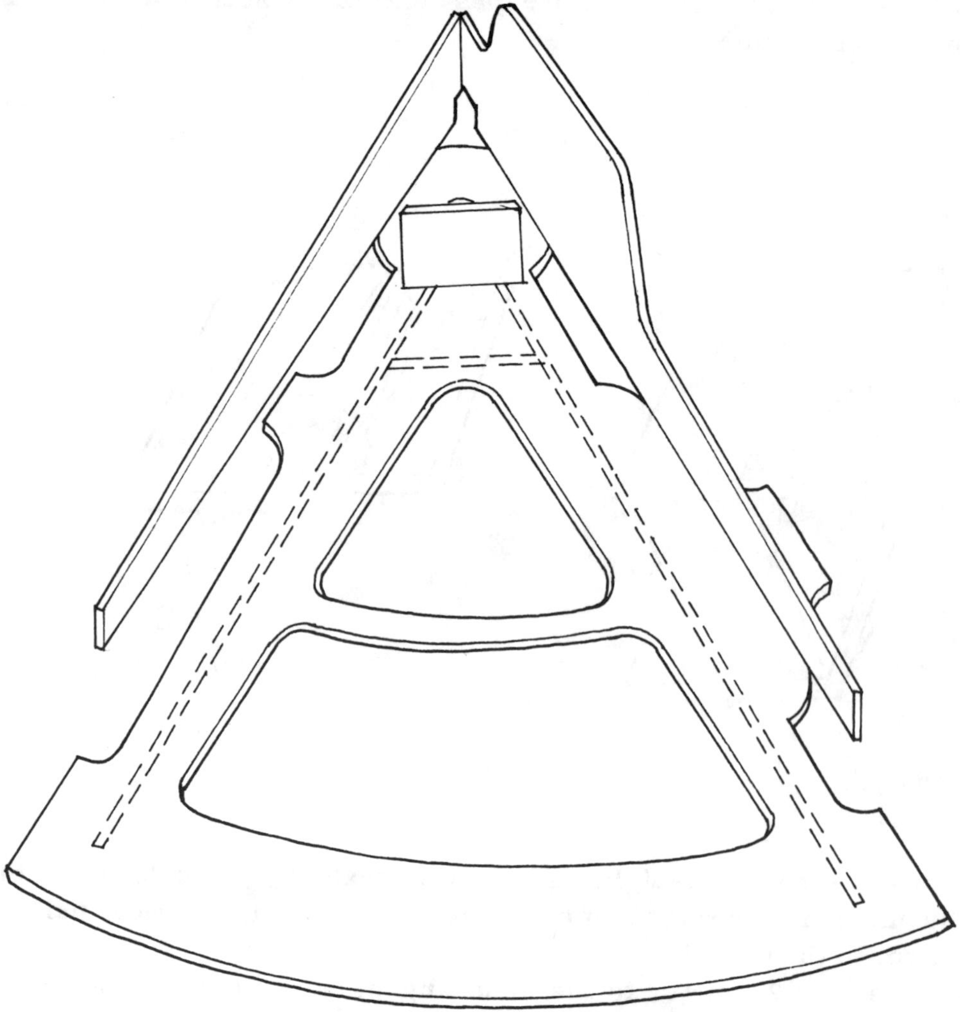

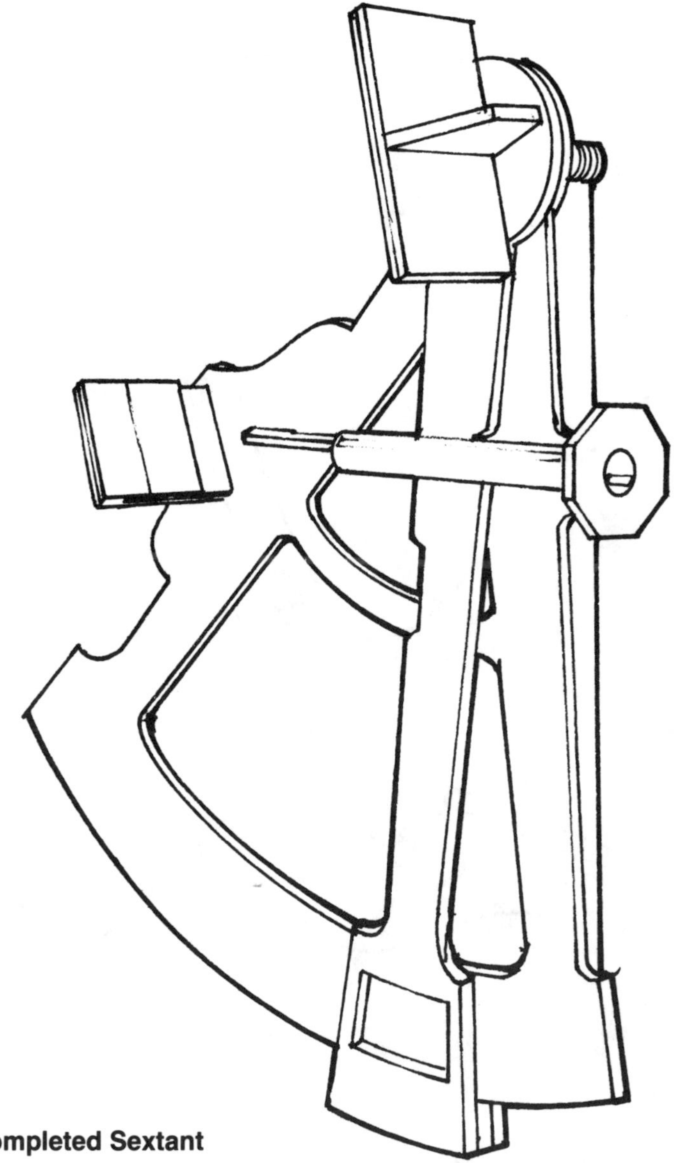

Completed Sextant

SEXTANT PATTERN

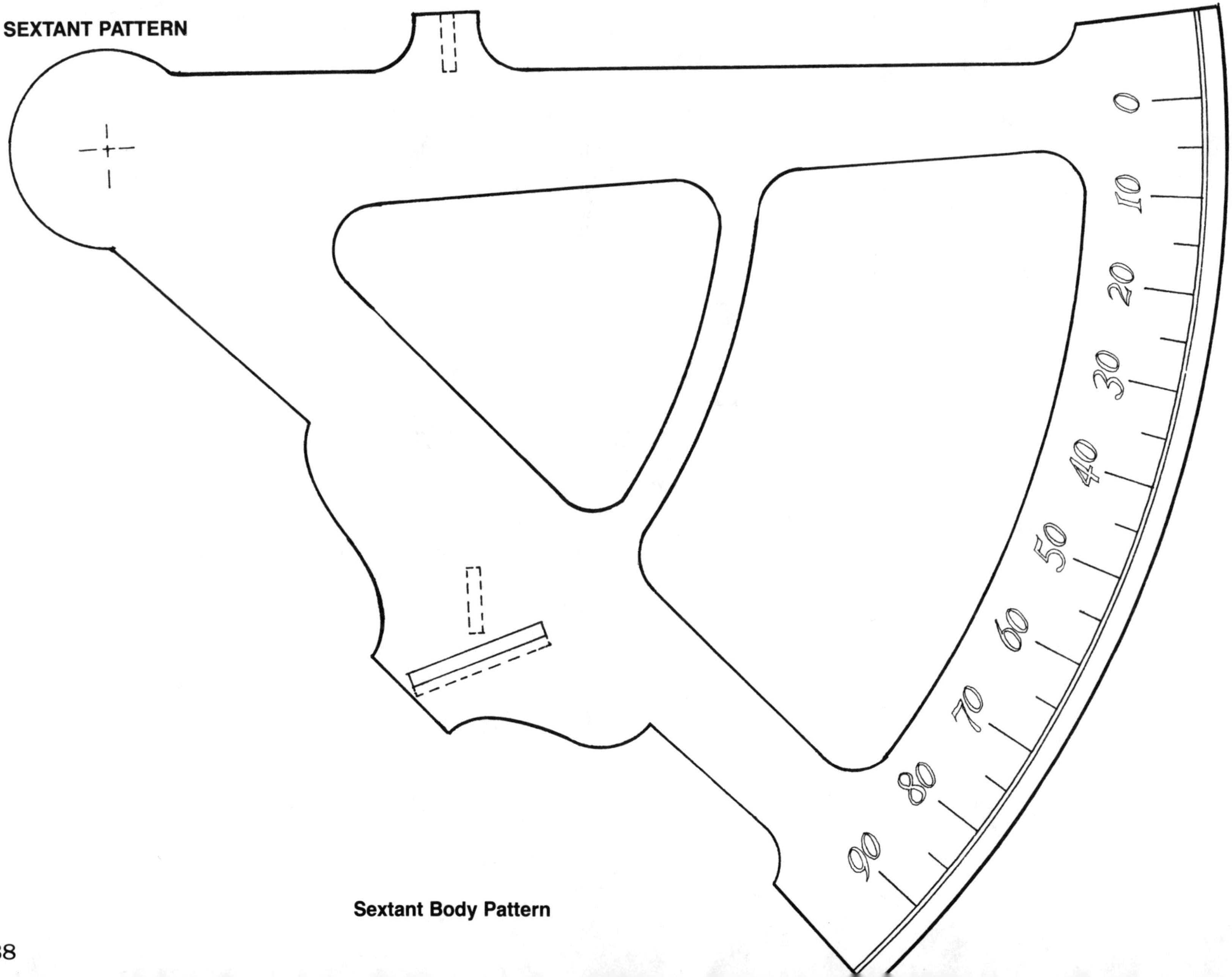

Sextant Body Pattern

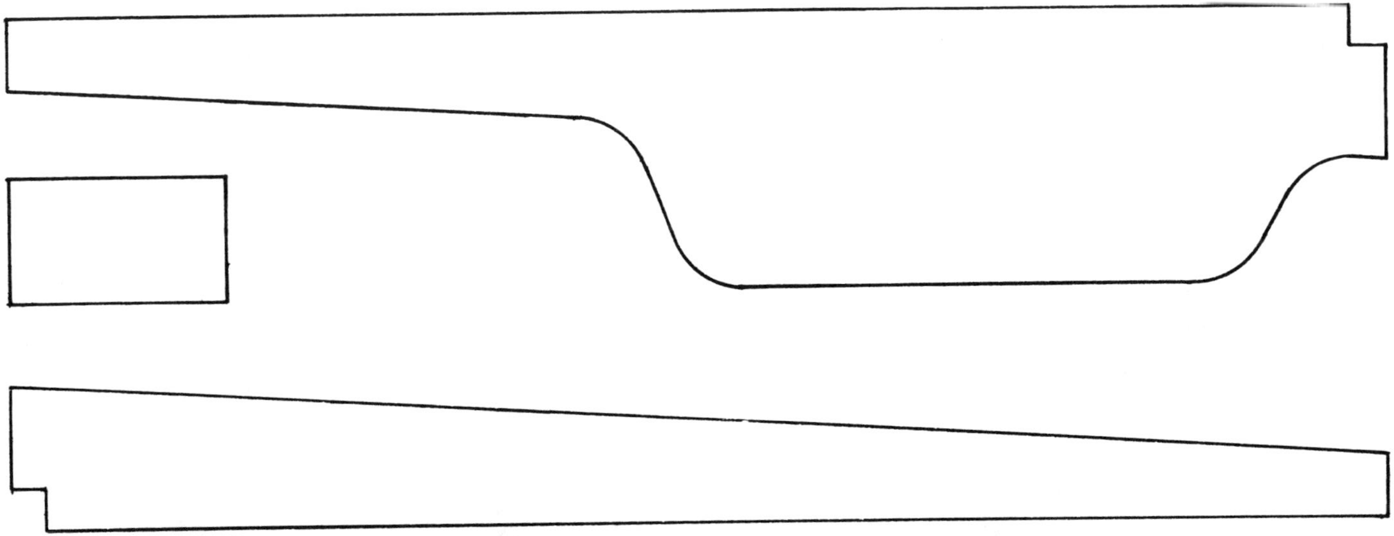

Sighting Support Pieces

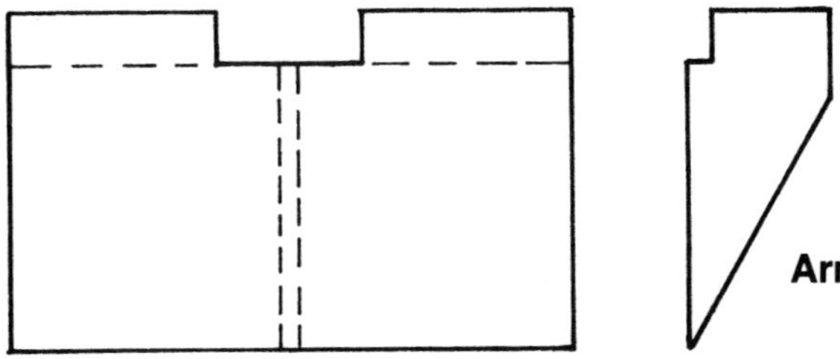

Arm

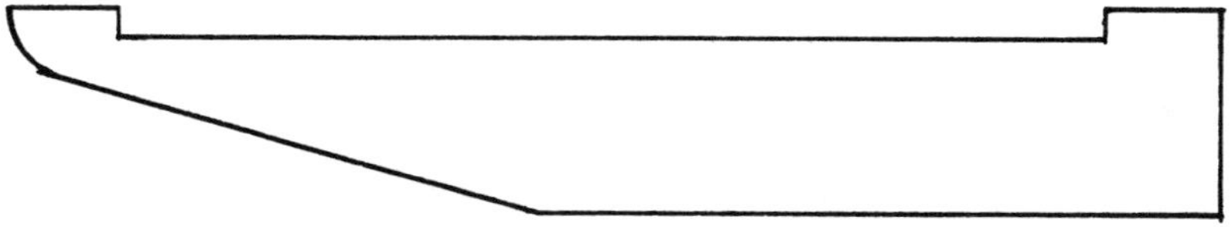

Sighting Mount

Body

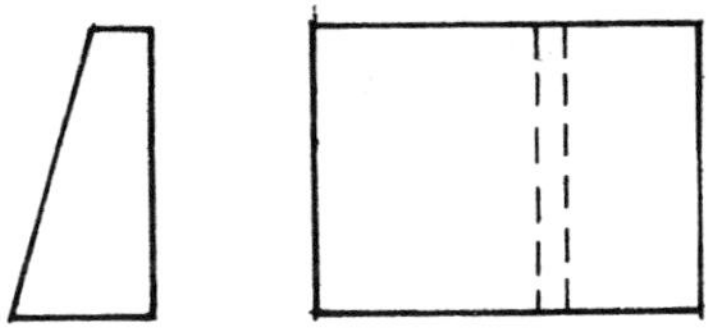

Mirror Mounts

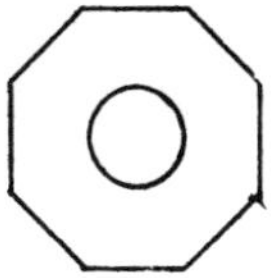

Eyepiece

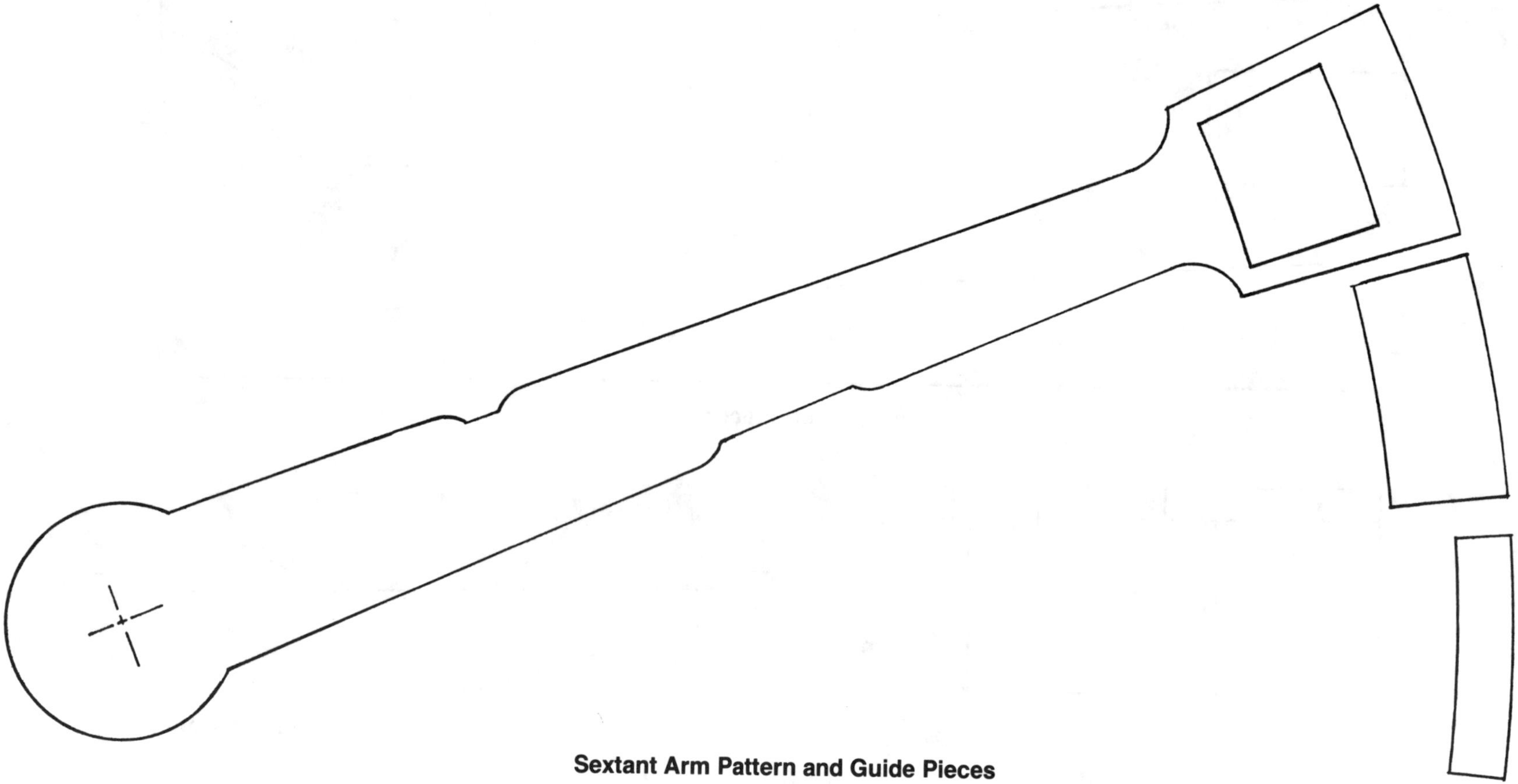

Sextant Arm Pattern and Guide Pieces

Before you can use your sextant, you must set its horizon line. Complete the following steps to do this:

1. Sight the horizon through the sighting tube and the clear half of the horizon mirror as shown by the line segment E, HM, H^1 in the illustration.

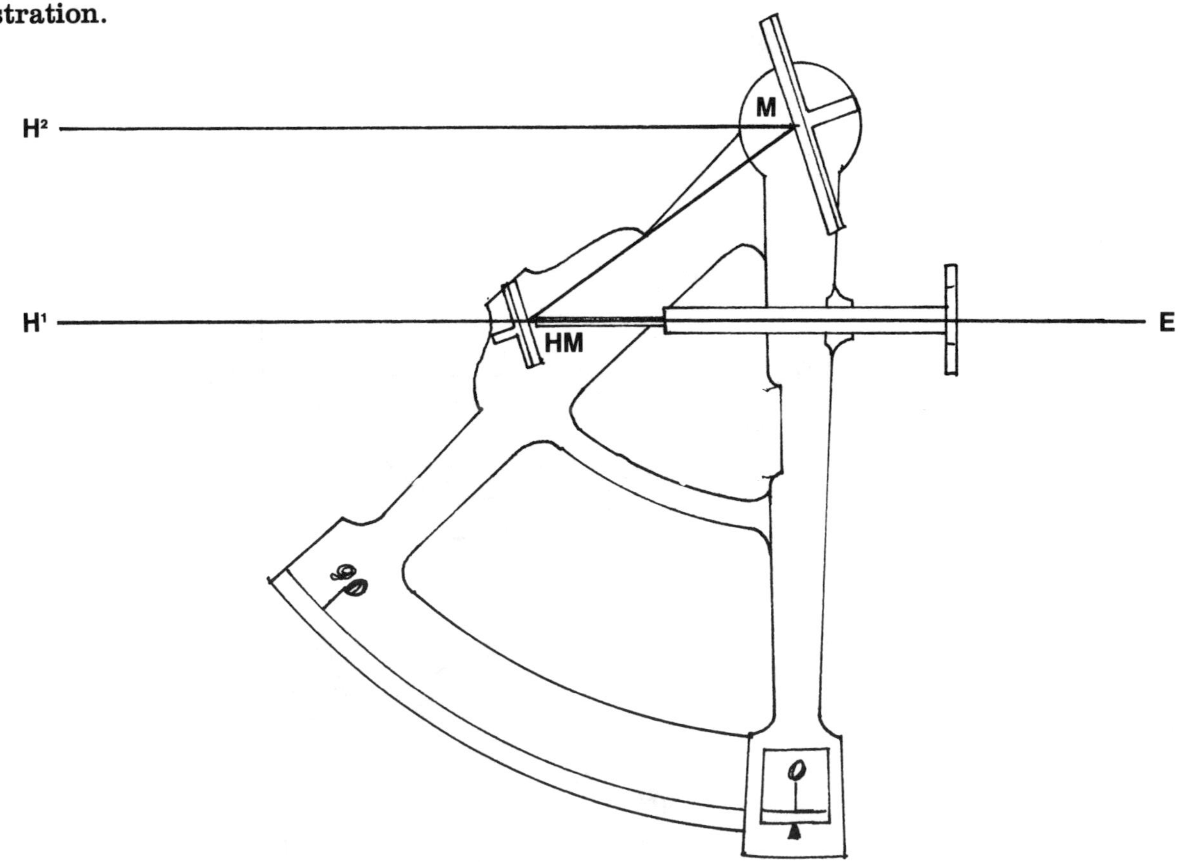

2. Slide the arm until the upper mirror sights the horizon and reflects its image on the mirrored half of the fixed mirror, completing the horizon line. Mark on the arm that lines up with the zero on the degree scale.

In order to make an astronomical observation with your sextant, follow the steps listed below:

1. Slide the arm until the heavenly body reflects its image on the mirrored half of the fixed horizon mirror and lines up exactly with the horizon line.

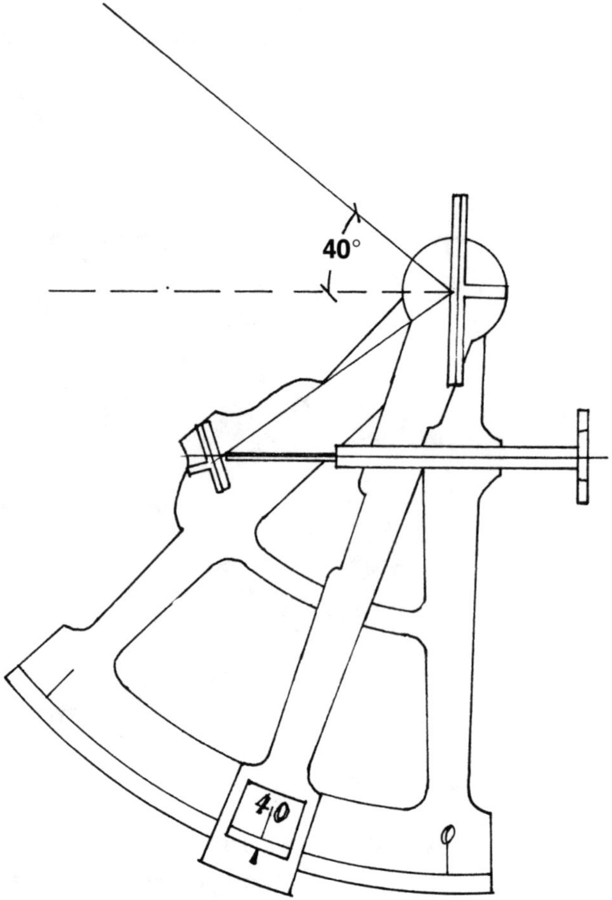

2. Read the number of degrees indicated by the arm on the degree scale. In the illustration, the degree reading is 40.
3. Note the day and time of day of the observation. If you check accurate star tables, such as the *Nautical Almanac* or *Air Almanac*, you will find the location on Earth that a specific star can be seen on a given day and at a specific time and altitude.

You may want to make two or more observations of different celestial bodies in the sky from the same location, or observations of the same object at different times to determine more exactly the latitude and longitude of your location on Earth.

The development of portable astronomical instruments such as the sextant was important to astronomers as well as sailors. Tycho Brahe was one of the astronomers who used portable instruments in his scientific observations. Brahe's various observations helped make him an important link between the theories of Copernicus and later

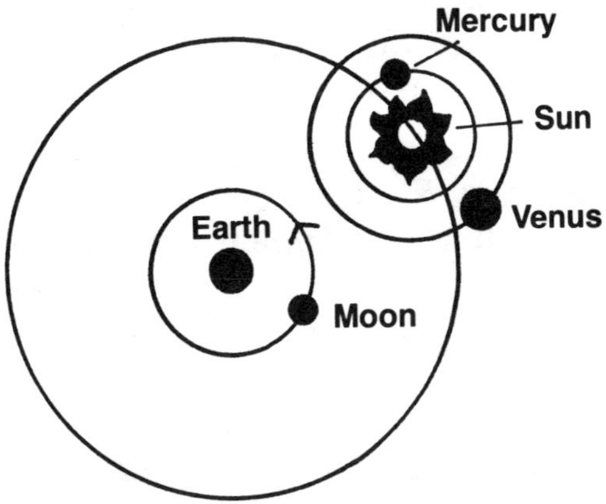

Tycho Brahe's system of the universe

scientists and astronomers. He did not believe that the heliocentric universe of Copernicus was possible based upon the known laws of **physics**. Instead, he still believed in a modified geocentric solar system in which the planets revolved around the sun and the sun in turn circled around the Earth.

PHYSICS

The science concerned with matter and energy, and the relationship between them.

After living and working at his island castle observatory for over 20 years, Brahe began to have a difficult time getting along with the new Danish king, Christian IV. The king cut his allowance, so Brahe left Denmark and traveled around Europe for two years. By 1599 he was in Prague, the largest city in Czechoslovakia. It was in Prague, a year later, that he met a young mathematician, Johannes Kepler, who had come to the city to work with him.

Kepler became Brahe's assistant in Prague, at the small castle Brahe was using as an observatory. Kepler had copies of Copernicus' book *On the Revolutions of Heavenly Spheres* and Brahe's catalog of the positions of stars and planets. For years he observed the heavens and filled over 900 pages with his own observations and calculations. Eventually, Kepler published his own set of star tables, which combined his observations and Brahe's, and which became the basic manual for calendar makers and astronomers for the next hundred years.

Johannes Kepler (1571-1630) *Culver Pictures*

Kepler was born in Württenberg, Germany in 1571. Although his grandfather was a knight, Kepler grew up in poverty and remained poor throughout his life as an astronomer. As a student, his favorite subject was mathematics, which he later taught. While a student, Kepler also had dreams of traveling through space and visiting the moon. In fact, he wrote *Somnium*, a book about space travel to the moon. It was probably the first science fiction story ever written. He described the dark frozen side of the moon, as well as the bright side. Kepler even thought creatures might live on the moon, much like the prehistoric reptiles that inhabited the Earth millions of years ago.

11/29

Marel —

Nyani brought these books in this a.m. — she asks that they be donated to the girls' Library.

Ginny

By the 1590s, Kepler was very much involved in astronomy. He was one of the first important astronomers to believe in Copernicus' theory of the solar system. He was most concerned with discovering the mysteries of the spaces between the different planets. Kepler finally decided that perfect geometric shapes, like the **tetrahedron** or the **hexahedron**, filled in these spaces. The ballshaped sun was at the center of a geometric universe. Even though it was very difficult for Kepler to show how his geometric model of the solar system worked, he used it as the basis for exploring three laws of planetary movement. He discovered that the movement of the planets around the sun was not circular but **elliptical**. The sun was the focus of the ellipse of each planet. Throughout his studies, Kepler insisted on practicing an astronomy that was based on mathematics and physics. In doing so, he was responsible for developing the science of the physics of heavenly bodies, or celestial physics.

Kepler spent most of his life as an astronomer working with inadequate instruments. He was myopic, or nearsighted, and sometimes even saw double, so astronomical observations were often very difficult for him. Kepler wrote letters to both Galileo Galilei, the famous Italian professor of mathematics, and Tycho Brahe asking them to make certain observations for him. Both of these men had much more accurate instruments than Kepler. By 1600, Kepler decided to go to Prague to work with Brahe and use some of his instruments. The two men had only a very brief time to work together before Brahe died in 1601. At that time Kepler was appointed to replace him as the Imperial Mathematician for Rudolf II, the Holy Roman Emperor.

As Imperial Mathematician, Kepler had to draw up **astrological** calendars and **horoscopes.** He also wrote books on astronomy and

TETRAHEDRON
A solid figure with four faces or sides.

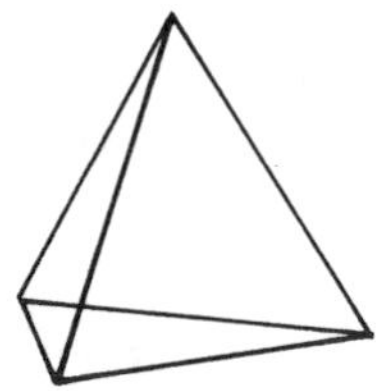

HEXAHEDRON
A solid figure with six faces.

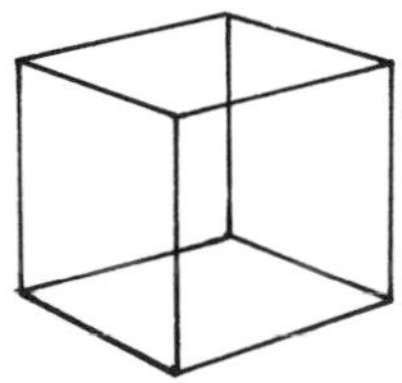

ELLIPTICAL
Shaped like an ellipse, an oval figure with both ends alike.

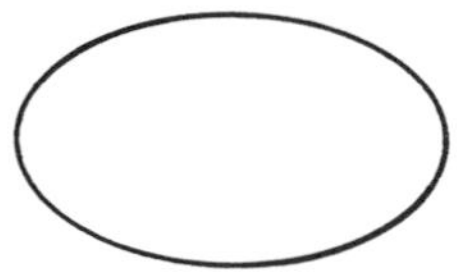

ASTROLOGICAL
Of astrology or the study of the planets and stars to determine their influence on people and events and to predict the future.

HOROSCOPE

A diagram that is used in telling fortunes according to the relative position of the planets at a particular time, especially at the person's hour of birth.

OPTICS

The scientific study of light and vision.

optics. In fact, in 1604, Kepler published a book which was the first to explain the basic principles of how human beings see and how eyeglasses help them to see better. He even suggested how these principles could be used to make a telescope. This was six years before Galileo first used his telescope to observe the stars.

The frontispiece designed by Kepler for his *Tabulae Rudolphinae* or star tables, published in 1627. It shows an imaginary observatory supported by columns. Some are in ruins and represent the ancient scientists. The two new columns in front represent Copernicus and Brahe. Brahe's island of Hven is depicted on the central slab. All kinds of astronomical instruments and tools are also illustrated. *Courtesy of the Dibner Library, Smithsonian Institution*

In 1611 another book on optics by Kepler, the *Dioptrice*, was published. In the work he examined why and how lenses **refract** light. His study founded the new science of refraction of lenses, or dioptrics, and established Kepler as the father of modern optics. It also provided the description of the telescope that is named after him.

REFRACT

To bend from a straight course.

The invention of optical lenses during the late Middle Ages made important changes in many people's lives. Probably sometime during the late 1200s, glassmakers discovered that glass discs could be used to make objects placed in front of them larger. At that time, the English monk Roger Bacon, not only described the use of such lenses, but he also suggested that they could be used to help people with poor eyesight see better. By the end of the 1200s, other authors tell of using spectacles or eyeglasses to help them read.

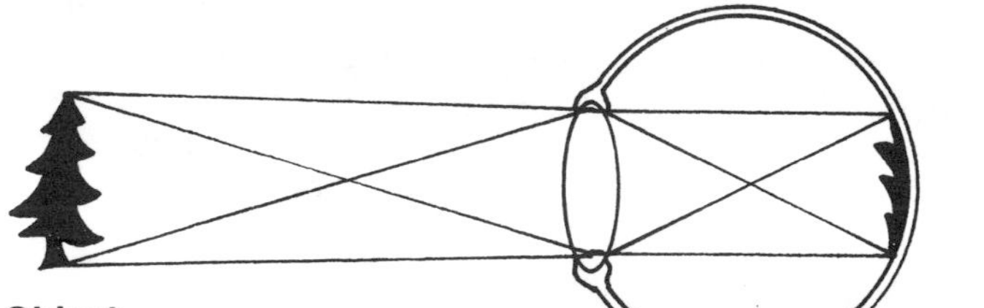

Human Eye

An illustration of a spectacle maker who made eyeglasses framed in leather or horn. From Jost Amman and Hans Sachs' *Eygentliche Beschreibung Aller Stands Auff Erden. . .*, (1568).

The reason that many people need eyeglasses is that their eyes have lenses that are poorly shaped, or are too thin, or too thick. Basically the human eye is a hollow sphere about one inch in diameter. Inside the eye is a **convex lens** made of transparent flexible tissue. Light enters the eye through the lens and focuses itself on the retina.

CONVEX LENS

An oval lens which has a surface that curves or bulges outward so that it is thicker in the middle than at the ends.

The retina is a delicate, light-sensitive membrane located at the back of the eye. It is connected to the brain by the optic nerve. When light strikes the retina, a chemical reaction takes place. An electrical impulse is generated in the optical nerve. This impulse is sent to the brain where it is interpreted as the shapes and colors of the things we see.

SUSPENSORY LIGAMENTS

Ligaments that support an organ or part of the body.

CILIARY MUSCLES

Microscopic hairlike muscles.

The lens is attached to the eye by **suspensory ligaments**. **Ciliary muscles** are attached to these ligaments. When these muscles are contracted they cause the lens in the eye to become thicker. When relaxed, the lens in the eye becomes thinner. By being able to change the shape of its lens, the eye can focus objects that are both near and far away on the retina.

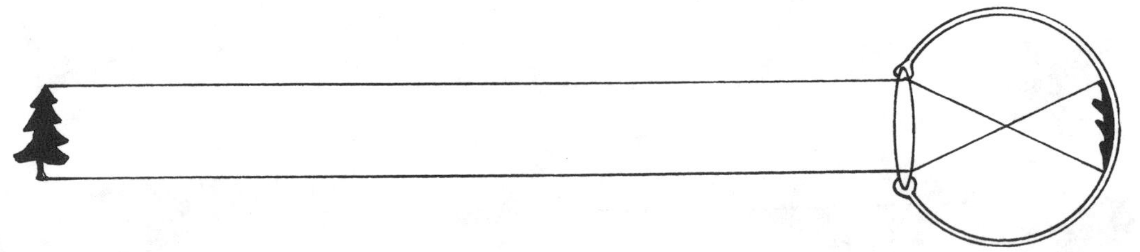

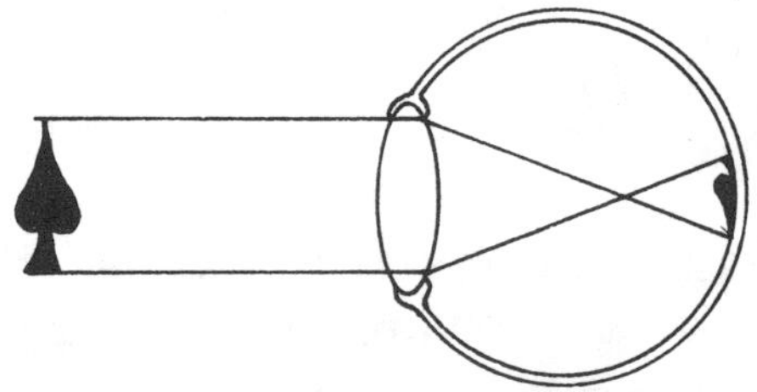

When people have eyeballs that are too long and lenses that are too thick, they will have difficulty seeing objects that are far away. This condition is called nearsightedness. People whose eyeballs are too short and whose lenses are too thin see faraway objects very clearly, but nearby objects with difficulty. This condition is called farsightedness. By placing a lens in front of the eye, these problems can be corrected. This is exactly what eyeglasses do to help people see better.

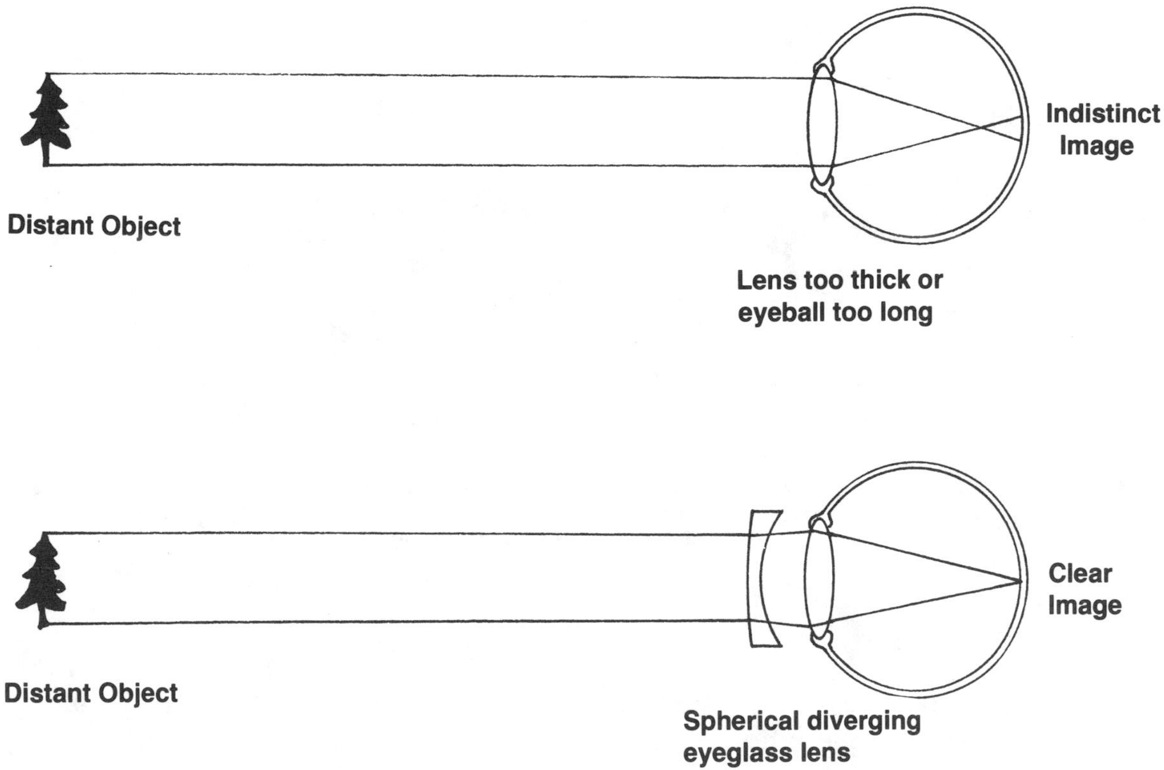

Eyeglass lens used to correct nearsightedness

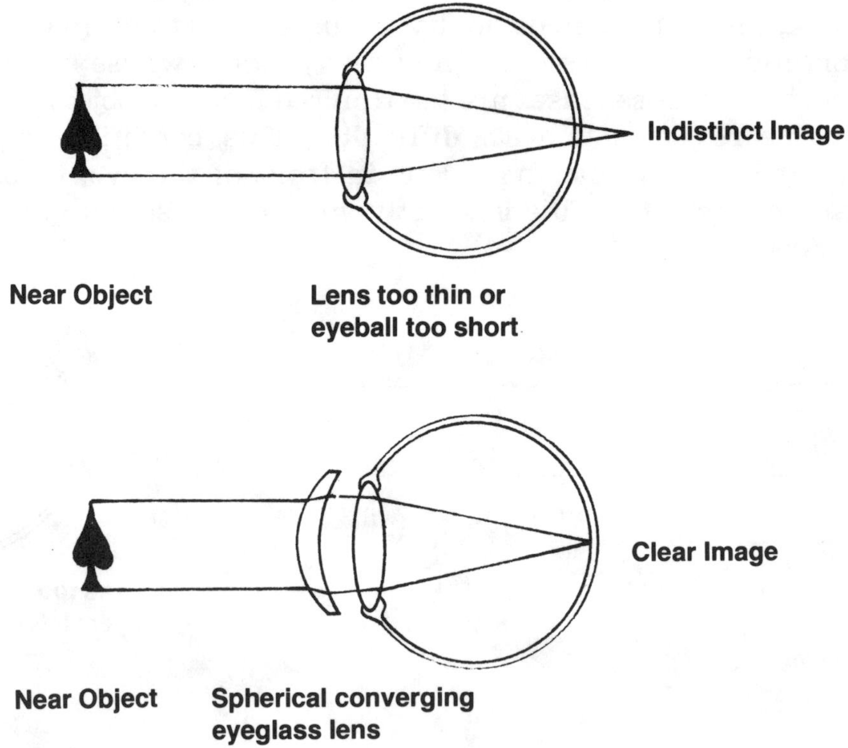

Eyeglass lens used to correct farsightedness.

The invention of spectacles had an extraordinary effect on the lives of many people who could suddenly see, read and write again. Eyeglasses made it possible for people to overcome their natural physical limitations and to better cope with the world in which they lived.

The invention of glass lenses, which could be used to observe and see objects, happened relatively late considering that the ancient Egyptians, as well as the Greeks and Romans, were aware of the fact that lens-shaped pieces of **rock crystal** could magnify and enlarge objects that were placed in front of them. These ancient people were also aware of the fact that water, if placed in a round glass or globe, can magnify an object and cause it to appear larger.

ROCK CRYSTAL

Crystals of pure quartz, a very hard mineral, that is colorless and transparent.

A water magnifier illustrated in a book from the 1800s.

You can see how water acts as a magnifier by making a simple water lens yourself. You will need the following materials:

- Pint or quart-sized cardboard or plastic ice cream container
- Sheet of clear plastic wrap
- Large rubber band
- Water

Cut out the bottom and 2 holes on each side of the container as illustrated.

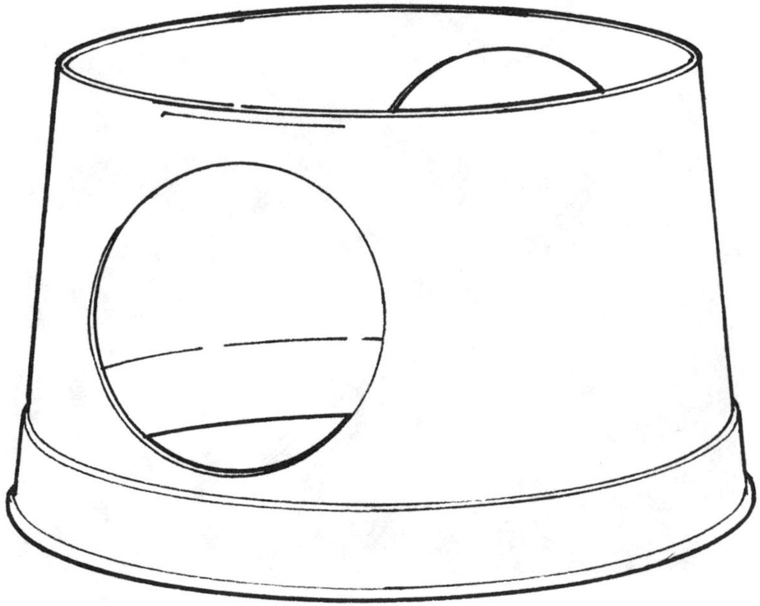

Place the clear plastic wrap loosely over the bottom of the container. Wrap the rubber band around the plastic and the bottom of the container so that it holds the plastic loosely in place. Pour a small amount of water onto the plastic wrap. The weight of the water should cause the plastic to sag and form a lens.

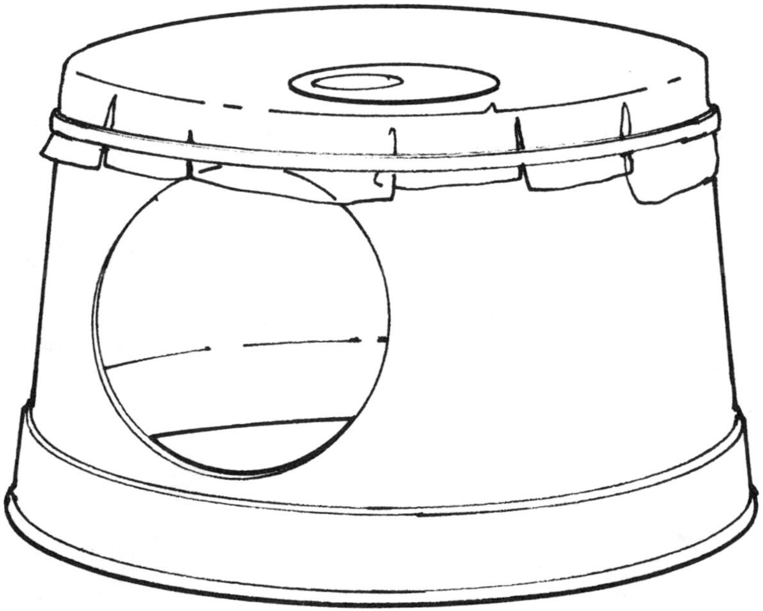

Place various objects under your water lens and see how they are magnified. You can change the shape of the lens by pouring different amounts of water onto the piece of plastic. The shape of the lens can also be changed by tightening or loosening the piece of plastic on top of the container.

The water magnifier works because the light reflected off of the object is bent by the lens and focused at a specific point. This point, where the reflected object concentrates its image, is known as the focal point. The image at the focal point is magnified because of the shape and substance of the water lens.

You can easily see how the focal point for a lens works by placing a magnifying glass in front of a light bulb. Use the magnifying glass to focus the inverted or upside-down image of the light bulb onto a piece of white paper. The magnifying glass will work just like the astronomical telescope described by Kepler.

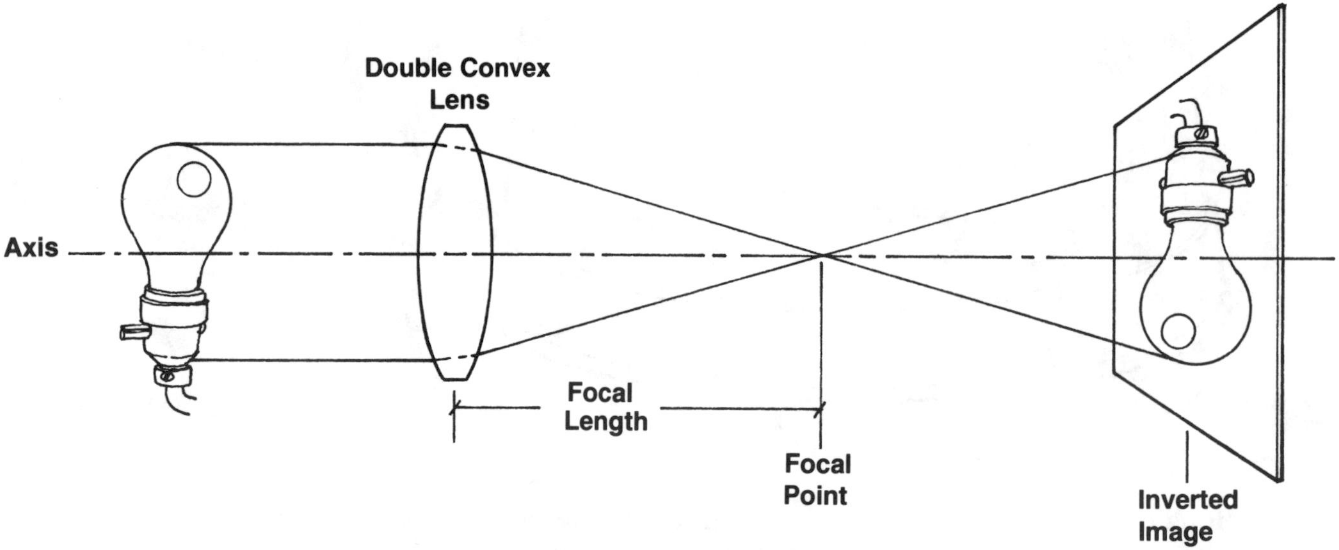

When the light goes through the lens, it is bent or refracted. You can see how light is bent by a substance such as water by setting a pencil into a glassful of water. Look straight down into the water. Place your finger on the side of the glass where the tip of the pencil appears to be. Then look at the side of the glass to check where your finger is placed and the real position of the tip of the pencil. Your finger is not marking the real position of the pencil because the light rays from the bottom of the glass are bent by the water which causes the pencil to appear to be higher in the glass than it really is.

Light is also refracted by the shape of lenses. Lenses may be convex, or **concave**, which is the opposite of convex. They can also be double concave or double convex; plano concave or plano convex; or concave meniscus or convex meniscus.

CONCAVE

Hollow and curved like the inner surface of a sphere or circle.

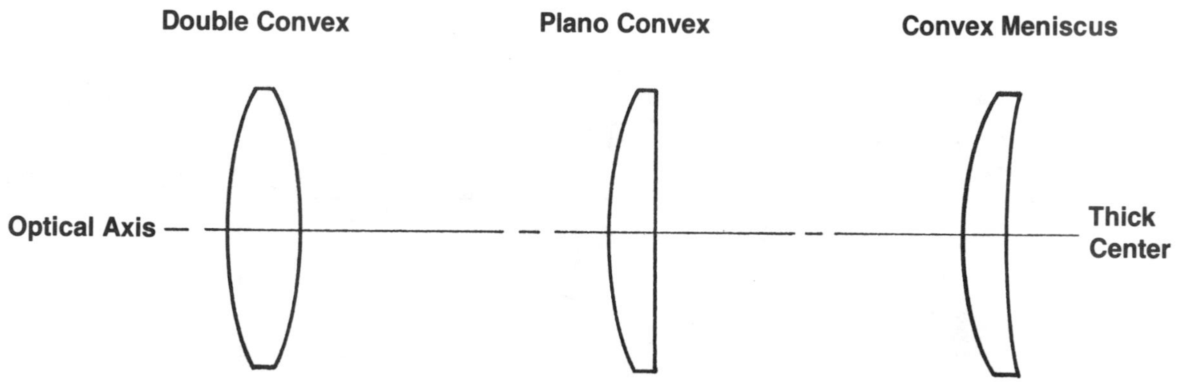

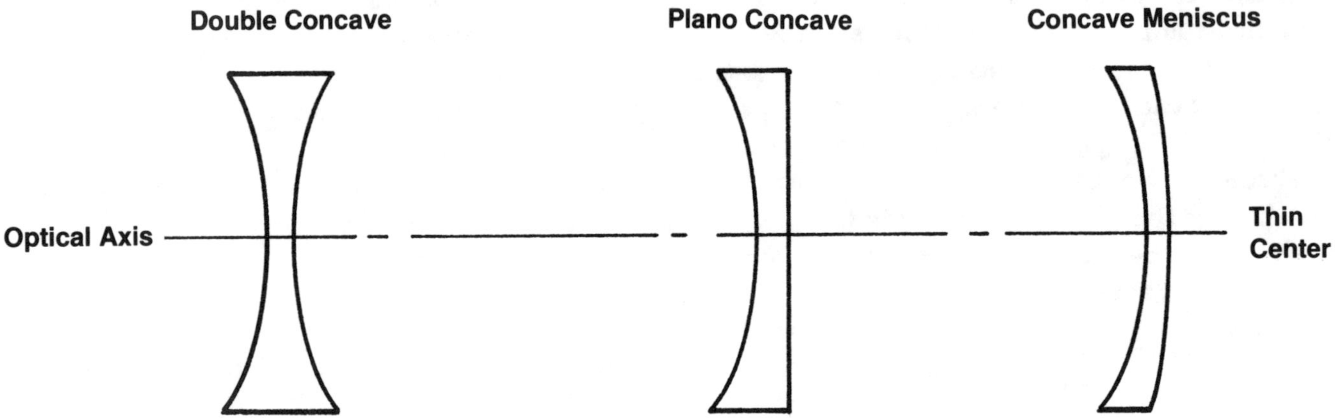

Convex and concave lenses refract light rays in different ways. When parallel rays of light pass through a convex or "positive" lens they are refracted twice, once on entering and again when they leave the lens. They converge, or meet again, at point "F" in the illustration, or the focal point.

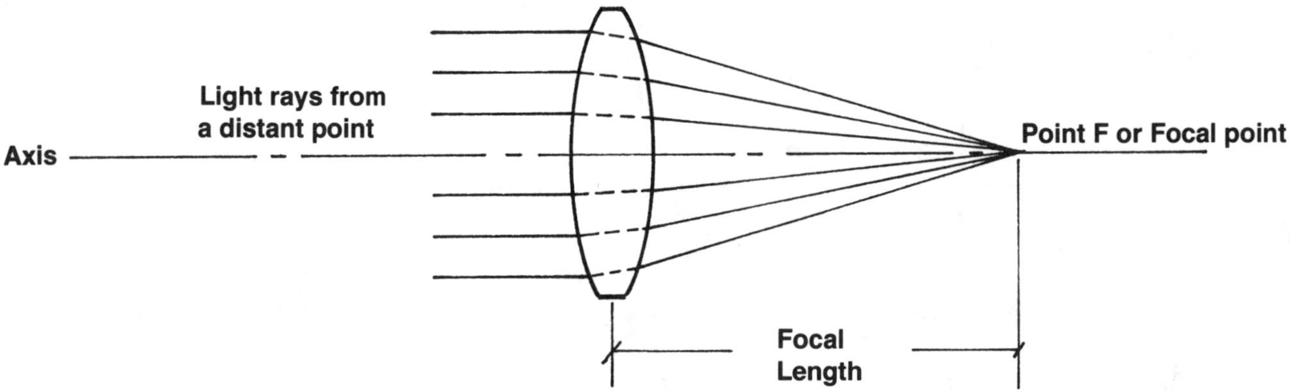

When parallel rays enter a concave or "negative" lens and then leave it, they diverge or branch outward.

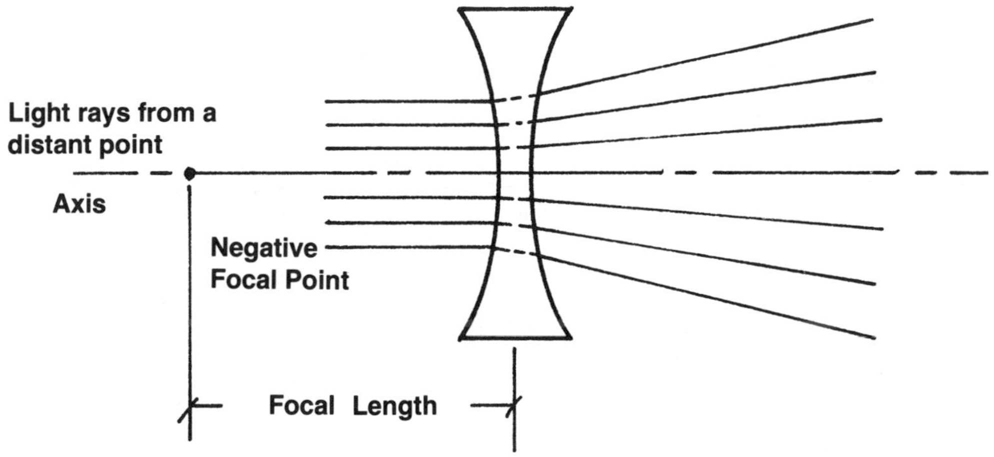

When these two types of lenses are combined with one another, they can form a telescope. In fact, the earliest known telescopes were made by using a convex lens and a concave lens. The convex lens is the objective lens where the rays of light from the object collect. They then form a small, bright image at the focal plane that the concave eyepiece lens can focus on. As this image leaves the other side of the double concave eyepiece it is magnified so that the real object being observed is greatly magnified for the eye.

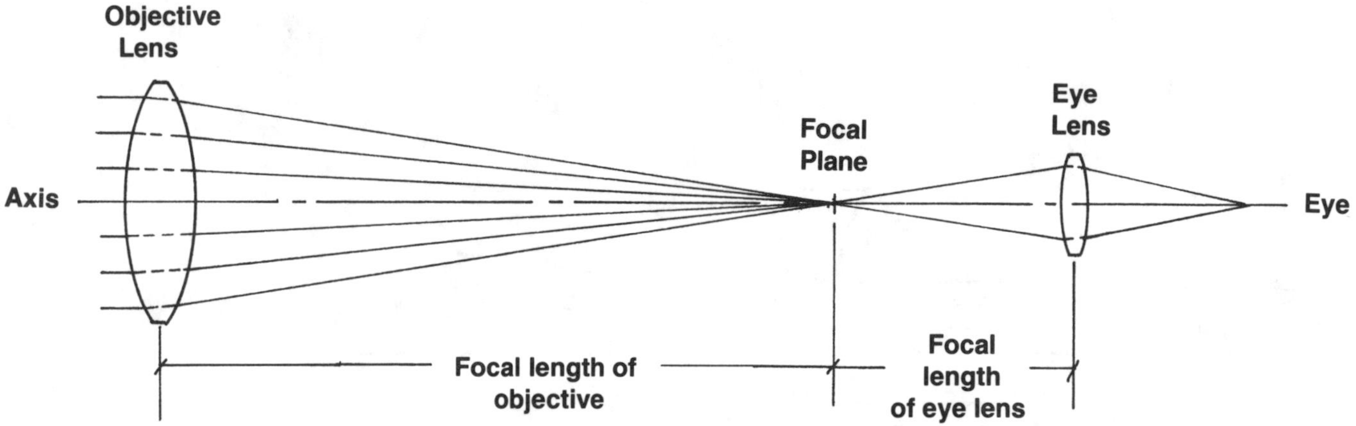

Diagram of a simple telescope

FOCAL LENGTH

The distance between the lens and the point where the light is focused, or the focal plane.

A telescope's magnifying power is determined by the **focal length** of the objective lens and of the eyepiece. The magnifying power is equal to the objective focal length divided by the eyepiece focal length:

$$\textbf{Magnification} = \frac{\textbf{focal length of objective}}{\textbf{focal length of eyepiece}}$$

For example, if the focal length of the objective is 12 inches and that of the eyepiece is 1 inch, then the magnifying power of the telescope is 12 ($12 \div 1 = 12$). The focal length of the objective remains the same. Changes in the focal length of the eyepiece, though, can change the magnifying power of the telescope as you are using it. A shorter focal length for the eyepiece increases the magnification.

The earliest-known telescope was made in 1608 by a Dutch lens grinder and spectacle maker, Jan Lippershey. He was denied a **patent** on the instrument though, because the Dutch government believed that other people also knew how to make telescopes. Lippershey's first telescope was probably about 1½ feet long and may have been made out of a tube of varnished, rolled paper. It used a convex lens as the objective lens and a concave lens as an eyepiece. Once his discovery became known, almost any spectacle maker could have constructed telescopes.

PATENT

A government document which gives a person or company the sole right to make, use, or sell a new invention for a certain number of years.

By 1609, telescopes were on sale in Paris. One had been brought to Venice, Italy, by a Frenchman who hoped to interest Italian sailors in using it for observations at sea. Many people also realized that telescopes could be used to observe enemy troop movements.

It was in June 1609, that the great Italian physicist, Galileo first heard about the new instrument. He reinvented it for himself and constructed a simple telescope out of a lead tube. Galileo was the first person to use the telescope as a tool of science. He was not so much interested in how the telescope worked as Kepler was, but more in what he could see with it.

Galileo's second telescope was an improved version made out of a rolled sheet of tinned iron plate. The edges were soldered and covered with crimson cotton. A convex lens was fitted into the tube as the objective lens. A concave lens was fitted into the opposite end for the eyepiece. Throughout his life, Galileo made about 100 different telescopes. The largest was probably 1¾ inches in diameter and magnified up to 20 times. Some of his telescopes were made out of wood covered with paper, or with leather decorated with gold.

Jan Lippershey (?-1619)

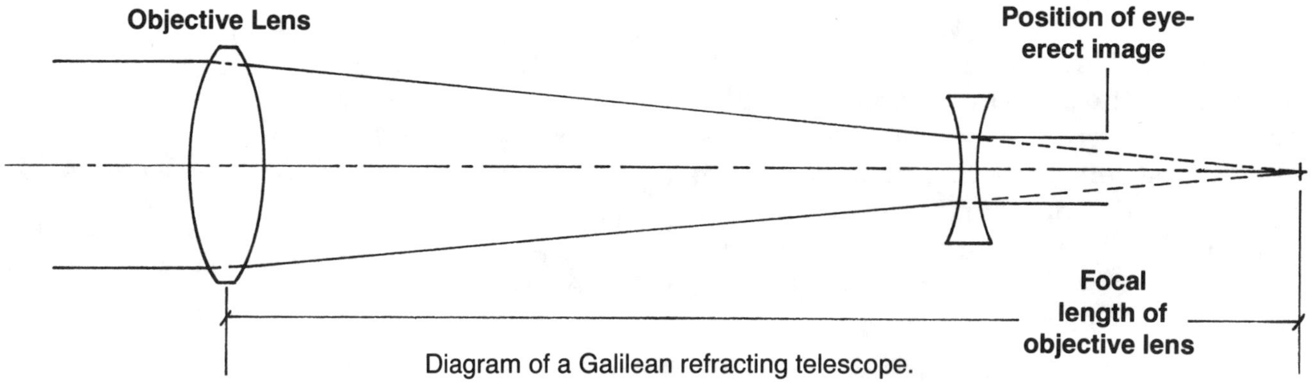

Diagram of a Galilean refracting telescope.

Galileo probably got the lenses for his first telescope from glassmakers in Murano or Venice. But the lens-grinding equipment available at that time was very crude. Not until the mid-1600s were major improvements made in this equipment by a young Florentine lens grinder, Ippolito Francini. Francini's improved machine tools produced lenses accurate enough that Galileo began to use them for his instruments. Another Italian, Guiseppi Campani, invented a **lathe**, at the end of the 1600s, which made it possible to grind polished lenses directly from the glass instead of from molded discs as Francini had done. Campani turned lens grinding into a science. His lenses and telescopes were considered the finest available in Europe at that time.

LATHE

A machine that holds pieces of metal and wood, and turns them rapidly against a cutting tool which shapes them.

Galileo was the first astronomer to use the telescope to look at the moon, the planets and the stars. His observations of the heavenly bodies helped him draw attention to the problems with Ptolemy's geocentric system of cosmology. The telescope allowed people to view the sky in a new way and helped Galileo promote his belief in Copernicus' heliocentric model of the universe.

Galileo was born in Pisa, Italy in 1564. As a child he is said to have enjoyed building his own toys. He also played the lute and organ and painted very well. While studying philosophy and medicine at the University of Pisa, Galileo discovered the law of the **pendulum**. When he was watching a great lamp swing from the ceiling of the cathedral, he noticed that each swing took the same amount of time, whether it was a large or small swing. This law makes it possible to use pendulums in large clocks. Its discovery was one of the first times a scientific law or principle was based upon observation and experimentation.

PENDULUM

A weight that is hung from a fixed point so that it is free to swing back and forth. The movements of a tall clock are often timed by a pendulum.

Galileo left his medical studies to do research in mathematics. By the time he was 25, he was a professor of mathematics at the University of Pisa. It was at this time that he made his second great discovery—the law of falling bodies. By dropping two unequal weights at the same time from the same height, Galileo proved that gravity pulls both bodies at the same speed, regardless of their weight, so that they hit the Earth at the same time. Most people opposed his new law, though, because they still believed in Aristotle's theory that heavier bodies fall faster.

Galileo Galilei (1564-1642) *Culver Pictures*

Because of opposition to his discovery, Galileo was forced to leave Pisa and became a professor of mathematics at the University of Padua. It was at Padua, in 1604, that he made his first important astronomical discovery. Galileo predicted that the new star which had appeared, a **nova**, would be visible for a while and then would fade away. This was a very daring prediction in the days when most people still thought of the universe as unchanging.

NOVA

A star that suddenly becomes brighter and then gradually fades back to its normal brightness.

Galileo's prediction about the nova was correct. Once he had built his first telescope, Galileo used it to help him find further evidence to support his belief in the Copernican system of the universe. For example, he discovered that the moon was not a perfect sphere as everyone thought, but that its surface had mountains and valleys just like Earth.

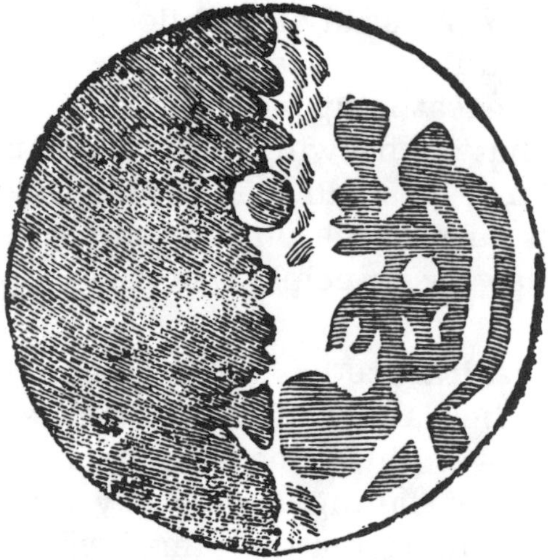

The features of the moon first seen by Galileo through his telescope. From an illustration in a 1655 edition of his book *Siderius Nuncius* (The Sidereal Messenger) printed in 1610. Galileo also described his introduction to the telescope in this book.

JESUIT

Member of the Society of Jesus, Roman Catholic religious order of men that was founded by Saint Ignatius Loyola in 1534.

SUNSPOTS

Any of the dark spots or regions of lower than average temperature that appear in groups on the surface of the sun, usually in 11-year cycles. When sunspots appear, disturbances in the Earth's magnetic fields often occur.

Galileo also made important observations of the sun. In 1612, the **Jesuit** astronomer Christoph Scheiner published a book on **sunspots**. Because he was interested in their opinion of his work, Scheiner sent copies to both Kepler and Galileo. Galileo responded by saying that he had known about sunspots for some time. Actually he probably did not know about sunspots before he received Scheiner's gift. Galileo's interpretation of what sunspots were, however, was correct when Scheiner's was not. Scheiner argued that the sunspots were bodies passing over the surface of the sun. Galileo said that they were part of the sun's surface.

Galileo was eventually able to prove his theory by systematically recording the position of various sunspots that appeared across the surface of the sun. Because of these observations, he was also able to determine that the sun rotated in much the same way that the Earth did.

Scheiner and Galileo used their telescopes to make their observations of the sun. They could not look at the sun directly with them, however. This would have blinded them. When you use your own telescopes, **NEVER LOOK AT THE SUN WITH THEM**. If you are interested in observing sunspots, take a pair of binoculars outside. Mount them on a laboratory stand or place them on some other stand that prevents them from moving.

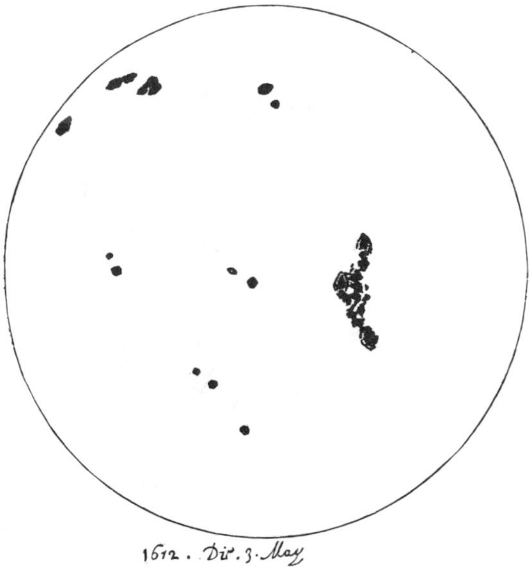

Galileo's drawing of sunspots

Position the binoculars in such a way that the sun will shine through them onto a piece of white cardboard. When you move the cardboard back and forth, you should be able to bring the image of the sun into focus. You should also be able to see small dark spots on the sun's image. These are sunspots.

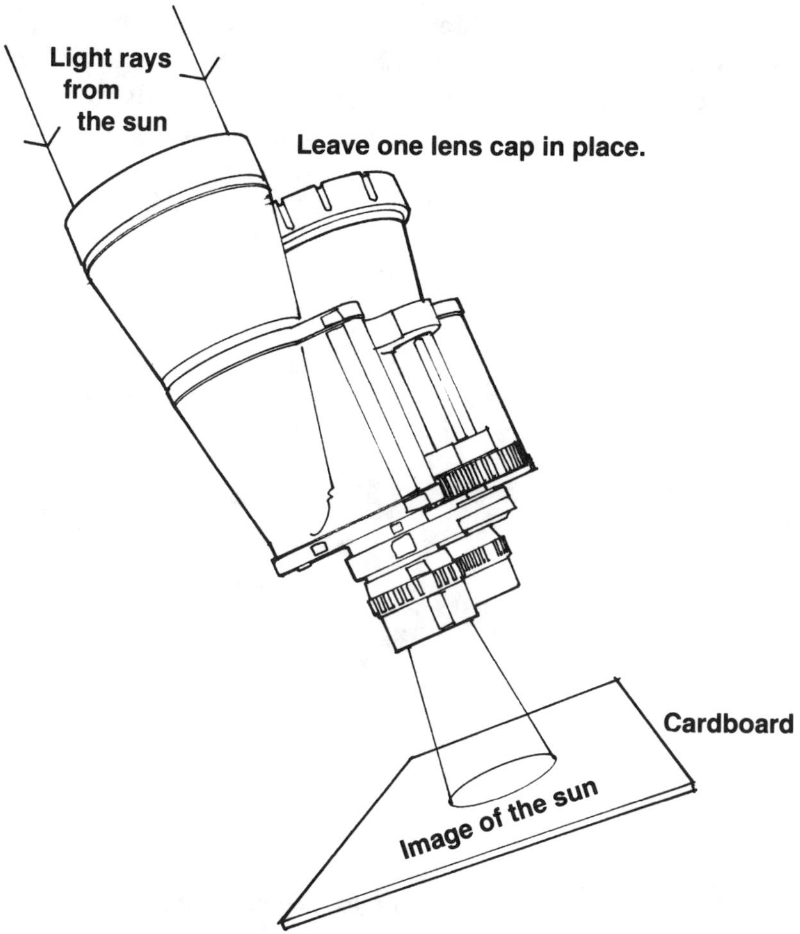

Trace the image of the sun and the sunspots on a piece of paper. Repeat the same observations at exactly the same time of day, with your equipment in exactly the same place, for four or five days. Trace the image of the sun and its spots each day on the same piece of paper or cardboard. You will see that the sunspots are moving across the surface of the sun. If you observe the sun long enough, you should also be able to record its rotation.

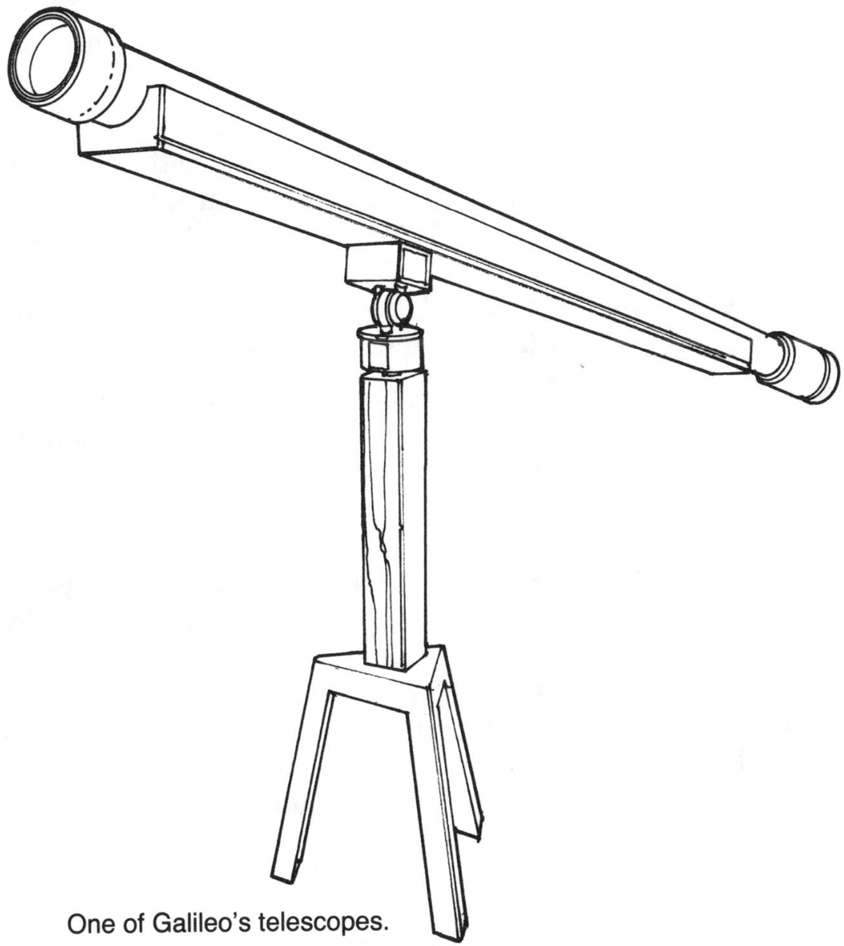

One of Galileo's telescopes.

SATELLITES

Small bodies that orbit around a planet. The moon is a satellite of the Earth.

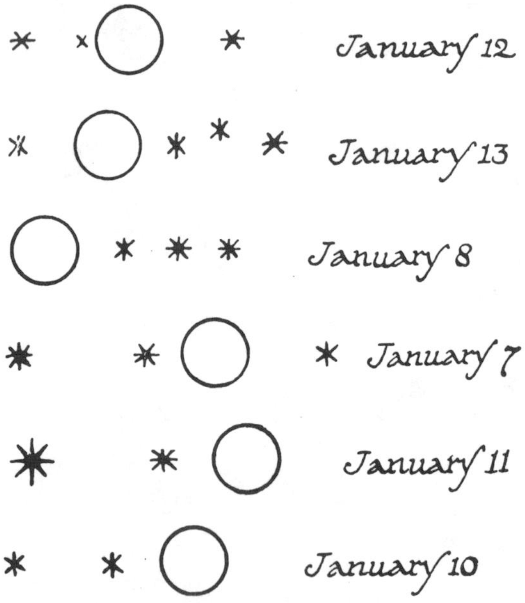

A diagram of the type of notations Galileo made when he was observing Jupiter's moons.

HOLY INQUISITION

A court established by the Roman Catholic Church in the 1200s to discover and punish beliefs different from the accepted beliefs of the church.

When he looked through his telescope at the planet Jupiter in January 1610, Galileo saw three **satellites** circling the planet. At first he thought they were small stars that were in his line of vision. But on the next night they were still there, except that all three had moved to the west side of the planet. Galileo began to watch the sky each night. He kept track of his observations by drawing the position of the planet and the three stars. His observations eventually established that there were actually four, not three, moons revolving around Jupiter.

Galileo believed that since Jupiter had four moons orbiting it and the Earth had one moon revolving around it, that the Earth and other planets could also be orbiting around the sun. This was the theory believed by Copernicus and Kepler. Galileo's telescopic observations provided new evidence to support their heliocentric theory of the solar system. He even observed that the sun itself experienced change as evidenced by the sunspots he saw through his telescope.

Galileo's observations and discoveries brought him fame throughout Europe. They also changed the way people viewed the solar system. Because many churchmen still followed the teachings of Aristotle, they saw Galileo's discoveries as threats. After moving to Florence, Galileo was warned to stop defending the Copernican system. When he published his great book, *A Dialogue on the World System* in 1632, he was summoned before the **Holy Inquisition**. Published in Italian (so more people could read it), the book outlined the differences between the system of Aristotle and Ptolemy and that of Copernicus. The book was banned and Galileo was tried. He was sentenced to an indefinite prison term which was served in confinement in his villa. He was also forced to deny or recant his belief in the heliocentric Copernican system of the universe.

Even though Galileo spent his last years writing on other scientific theories, the ideas of Copernicus, Brahe and Kepler did not die. Galileo had played an important role in confirming them through observations with his telescope and in publicizing them through his writings.

The title page of Galileo's *A Dialogue on the World System* shows the figures of Aristotle, Ptolemy and Copernicus holding the symbols of their systems of the universe. *Courtesy of the Dibner Library, Smithsonian Institution*

You can build a refracting telescope similar to the one Galileo first used to observe the moon and Jupiter. First you will need to learn how to roll your own paper tubes to make the lens/eyepiece systems. This is necessary since it is almost impossible to find pre-made tubes of the exact sizes required, especially since they must fit perfectly inside one another.

Making your own lens tubes and spacers to fit inside one another perfectly is really a simple job. You will need to remember the following points:

1. **Use a sheet of heavy paper (2-ply bristol paper works well. It does not buckle up when glue is applied.) An interesting note about paper is that it has grain, just like wood. Therefore, it will bend easier in one direction than another. To test your paper, cut a square out of one corner. Bend it both ways and you will see which way the grain runs. Mark the sheet and cut your strips across the grain so they will roll easier.**
2. **Pre-rolling, without glue will also help in starting the tube.**
3. **Use white glue. It is fast-drying and squeegees evenly.**
4. **Always make the tubes longer than needed, and trim them to the exact length required.**

MANDRELS

A spindle or rod around which material is shaped and supported.

DIAMETER

The measurement from one side through the center to the other.

The tools and materials required are simple:

- X-Acto knife (#11)
- Some small scraps of lightweight cardboard for squeegees
- Very find sandpaper (#600)
- 1 sheet heavy paper
- White glue
- Various **mandrels**

The mandrel is the most important item. Any straight, perfectly round rod, or tube of sufficient length and **diameter** will do. The mandrel forms the core around which the paper is rolled. If you cannot find a mandrel of the right diameter, you can increase the diameter of what you have by gluing paper around it. Paper accuracy is incredibly precise because each wrap around the mandrel increases its diameter by 2 paper thicknesses. Measurements can actually be made in thousandths of an inch!

To start rolling a tube, wrap the paper strip around the mandrel once. Try to wrap evenly. It may help to wrap 4 or 5 times around to be sure that it's even, then unwrap to just over one turn.

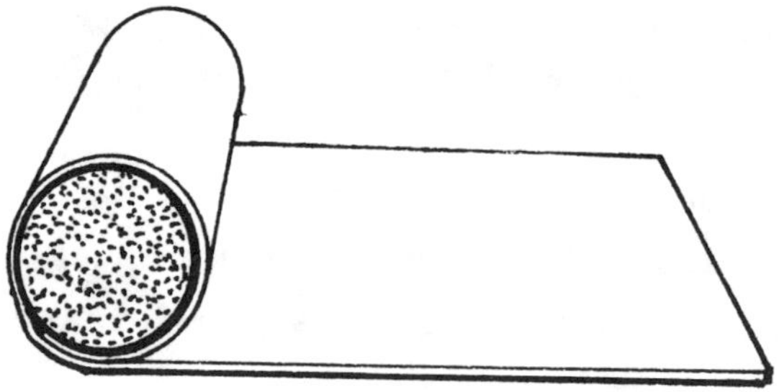

Mandrel

Squirt some glue onto a large scrap of paper. Using a squeegee, made from a ½-inch-wide strip of cardboard, pick up some glue and run a ½-inch-wide, thin layer along the flat area of paper adjacent to the tube.

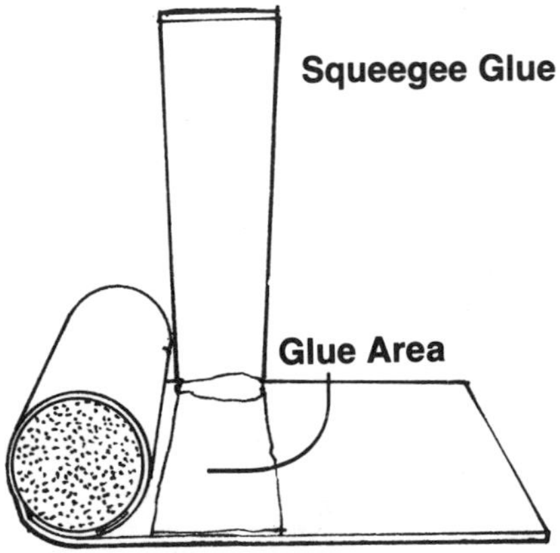

Slowly advance the roll over the glued area. Apply pressure as you go. The glue will be pressed ahead of the roll. If you over-roll, reverse direction until you return to the glued area. Continue gluing and rolling until the tube reaches the proper diameter. Finish off by carefully gluing the final flap and **burnishing** it down. A light sanding with #600 sandpaper is all that is needed to finish the surface.

In order to insure a perfect fit, always roll the smallest (inside spacer) tubes first. While they are still on the mandrel, roll the surrounding tubes directly over them. Be careful not to glue the tubes to one another! When all the tubes are made, slide them apart. Cut the spacer tubes to the proper lengths with an X-Acto knife, or fine-blade jeweler's saw. Sand the ends until they are smooth by rubbing the tubes, end down, over the sandpaper held on a flat surface. Remove any burrs, or rough spots, on the inside edges by lightly sanding with fine sandpaper rolled into a cone and twisted inside the end of the tube.

BURNISHING

To smooth, or polish, by rubbing with a smooth object.

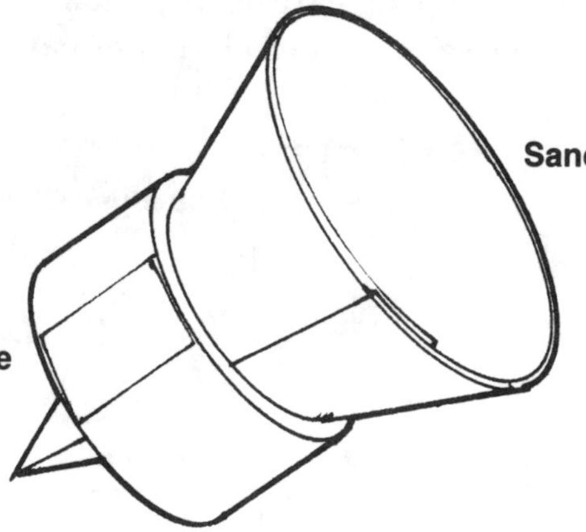

SHIMS

Thin pieces of wood, metal, or other materials that are used as a leveler or filler between materials or pieces of an object.

Once all of the tubes are trimmed and sanded smooth, spray them on the inside with quick-drying, flat-black paint.

It is a good idea to assemble all lens systems before gluing to check the fit of all of the components. Loose fits can be corrected with shims of Scotch tape. Apply tape to the inside of the tubes, building up their diameters until they fit properly.

The Galilean refractor telescope you are building is constructed in 2 basic units: the main tube and the focusing tube. Most of the required parts are standard plumbing devices available at hardware stores or plumbing supply houses. In order to build the telescope you will need the following materials:

Main tube—12″ PVC (polyvinyl chloride) extension pipe with threaded end connector, or slip nut

Focusing tube—8½″ section of 1″ PVC pipe and a 2½″ section of ¾″ PVC pipe

Eyepiece—rolled paper spacers and lens tubes and the part of the PVC unit called the "flexible riser" or tailpiece
Lenses—objective lens: one 44mm x 401mm plano convex lens
eye lens: one 14mm x 48mm double convex lens

Lenses are measured in millimeters and defined by two numbers: the first is the diameter of the lens and the second is the focal length. Therefore, 44mm x 401mm refers to a lens 44mm in diameter with a focal length of 401mm.

Lenses may be obtained from many sources, such as old cameras, or optical supply stores. Probably the most convenient source of accurate lenses for your telescope is:

Edmund Scientific Company
101 East Gloucester Pike
Barrington, New Jersey 08007

All of the lenses used in *Rediscovering Astronomy* can be ordered from this source. You may find it helpful to write to them asking for a copy of their free catalogue before you order any specific parts.

Polyvinyl chloride glue or cement (USE WITH CARE BECAUSE THIS GLUE IS VERY STRONG AND COULD GLUE YOUR FINGERS TOGETHER PERMANENTLY!)
Metric ruler
Flat-black spray paint
Very fine sandpaper
White glue

Using the PVC extension unit for the main tube of the telescope is a fast, economical and durable method. Just unscrew the threaded end section and paint the inside flat black.

Place the objective lens between the cap and the washer, curved side out and then replace the cap.

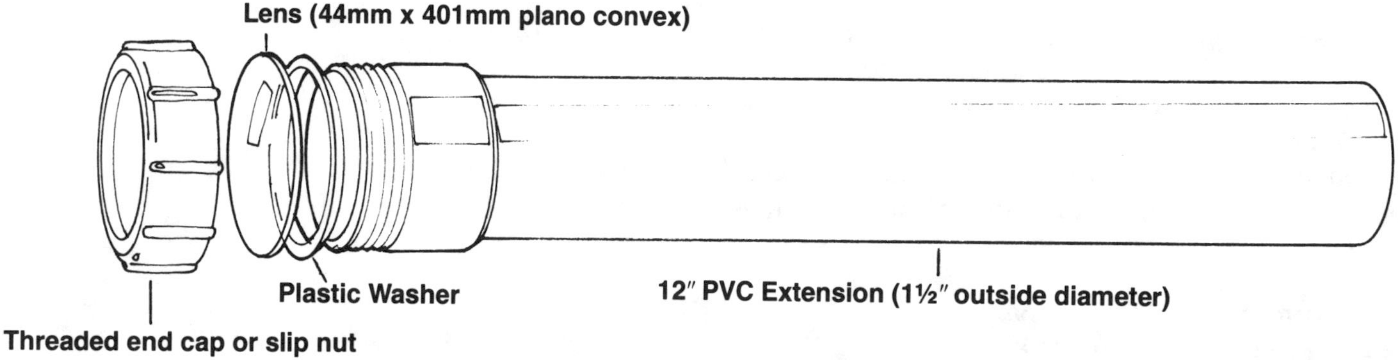

The focusing tube is made from the two pieces of PVC pipe. The smaller piece will probably need to be sanded down to accommodate a slip fit into the larger pipe. When the 2 pieces fit perfectly, use PVC cement to glue the smaller piece inside the larger one. Leave ⅝″ of the smaller tube extended.

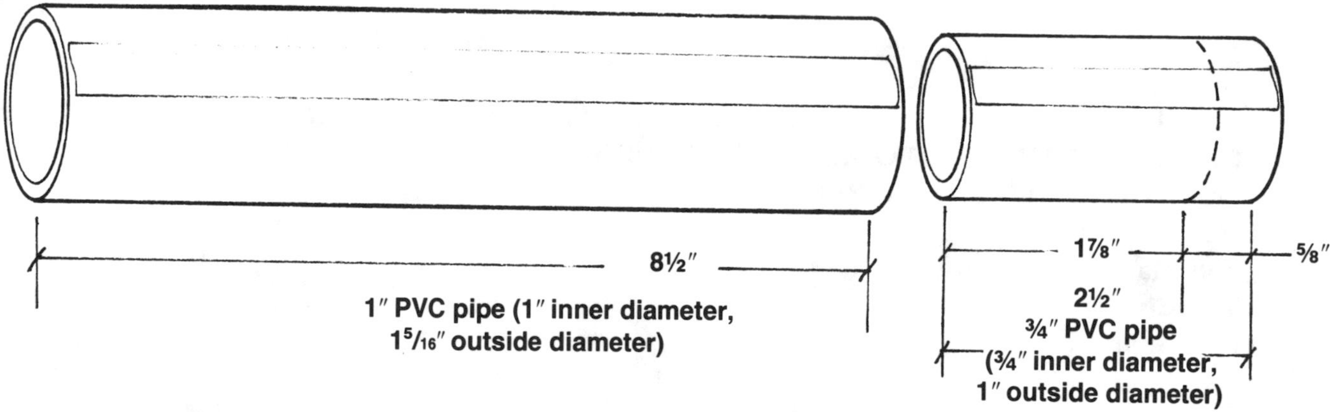

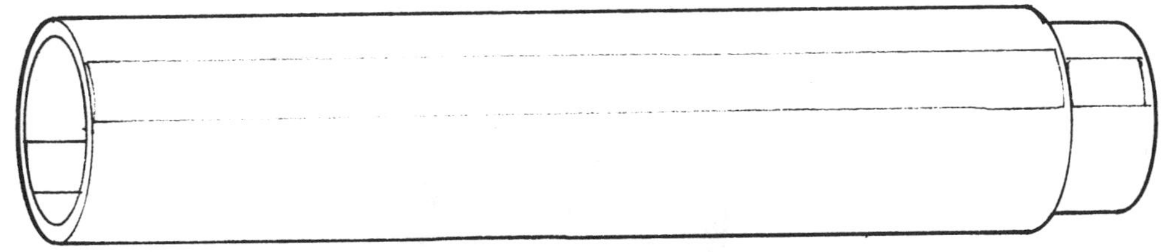

Roll and glue the eyepiece spacers around a 7/16″ mandrel to an outer diameter of 14mm (the same size as the eye lens diameter). Wrap the lens tube around the spacer tube to an outer diameter of ⅝″. Wrap another sleeve over the lens tube to an outer diameter of ¾″. Check to be sure that it will fit into the extended end of the focusing tube. Separate the tubes. Cut two 1″ sections from the small spacer tube. Cut the lens tube off true, or exactly, at 2¾″. Cut the outer spacer to a length of 2 3/16″. Sand all rough edges smooth and spray the insides of the spacers flat black.

The eye cup is trimmed from a molded plastic plumbing part called a "flexible riser." Cut the threads away from one end and trim away beyond the first rising section.

SLEEVE

Tube into which another tube or rod fits.

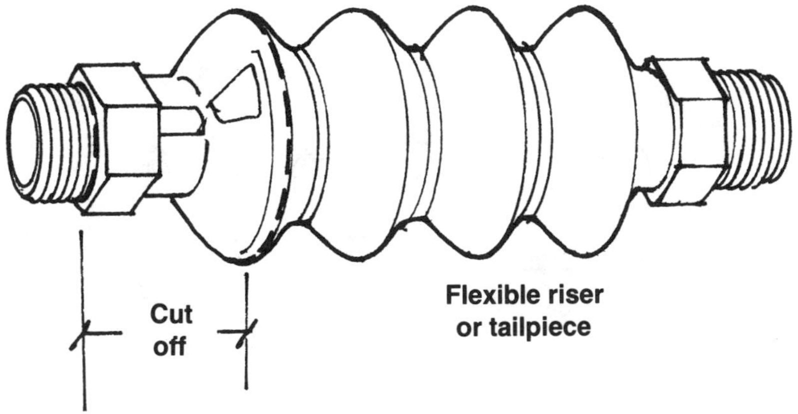

Lay out all of the finished eyepiece parts as shown below.

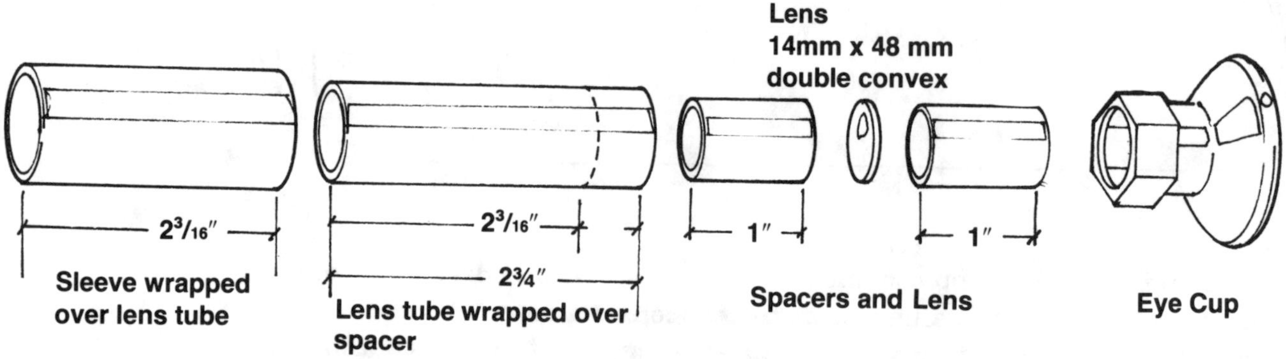

Glue the 2¾″ lens tube inside the 2³/16″ outer sleeve, allowing 9/16″ of the lens tube to extend. Glue one of the 1″ spacers flush into the extended end of the lens tube. Insert the 14mm x 48mm double convex lens into the other end of the tube, followed by the second 1″ spacer. Push the second spacer all the way into the lens tube until the lens is tightly trapped between the spacers. Do not glue the second spacer. It is left as a friction fit in order to make it possible to remove the lens in the future for cleaning or replacement. The eye cup is also a friction fit into the small end of the lens tube.

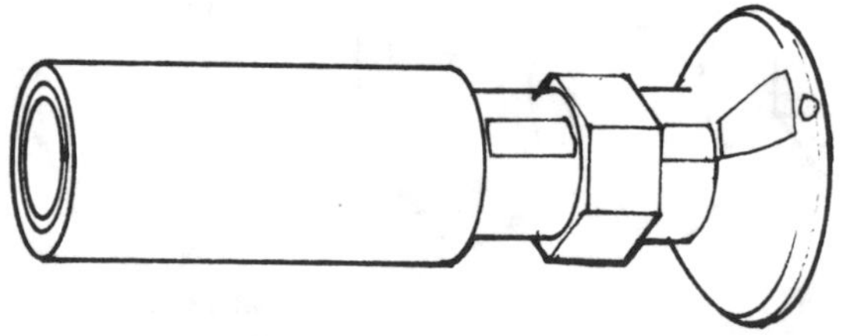

Assembled Eyepiece

The eyepiece unit is press-fitted into the small end of the focusing tube. The focusing tube should be wrapped with tape, to make a snug fit into the main tube possible.

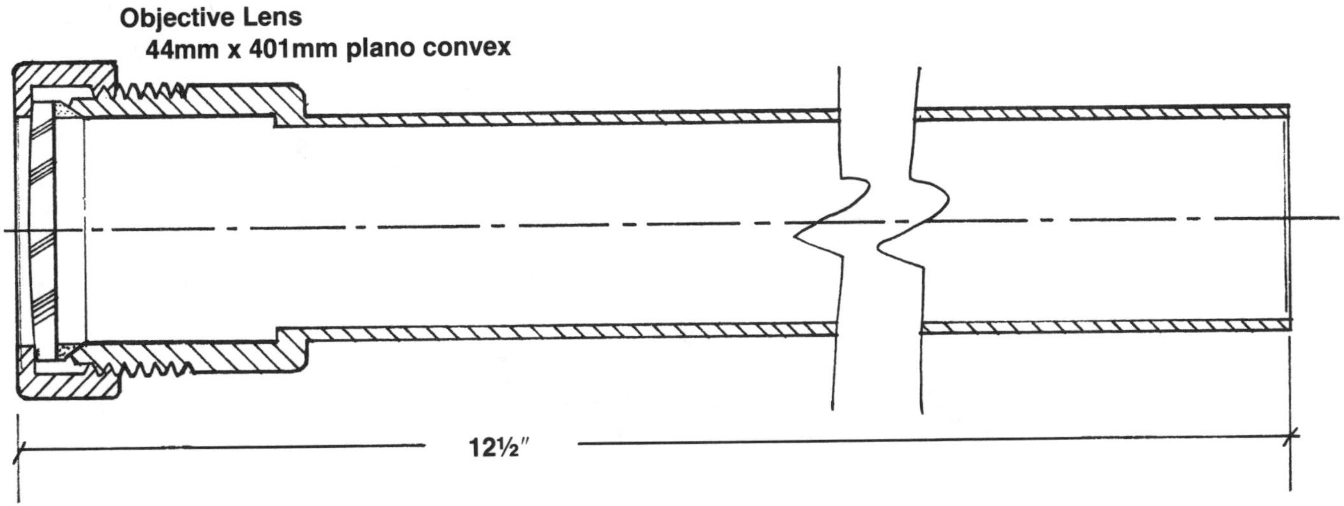

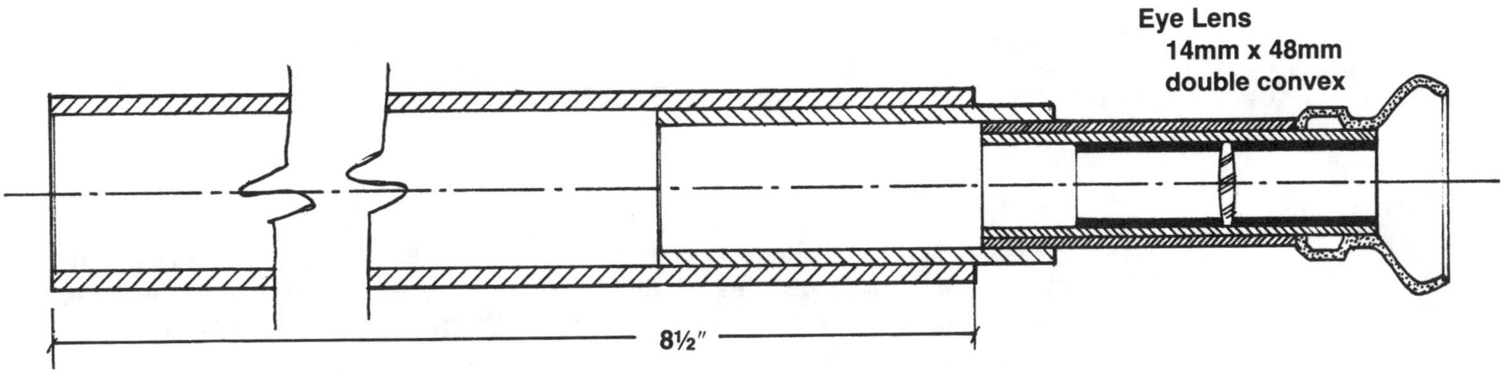

Slide the focusing tube in or out until the object you are observing comes into focus. Your completed Galilean telescope should look like this.

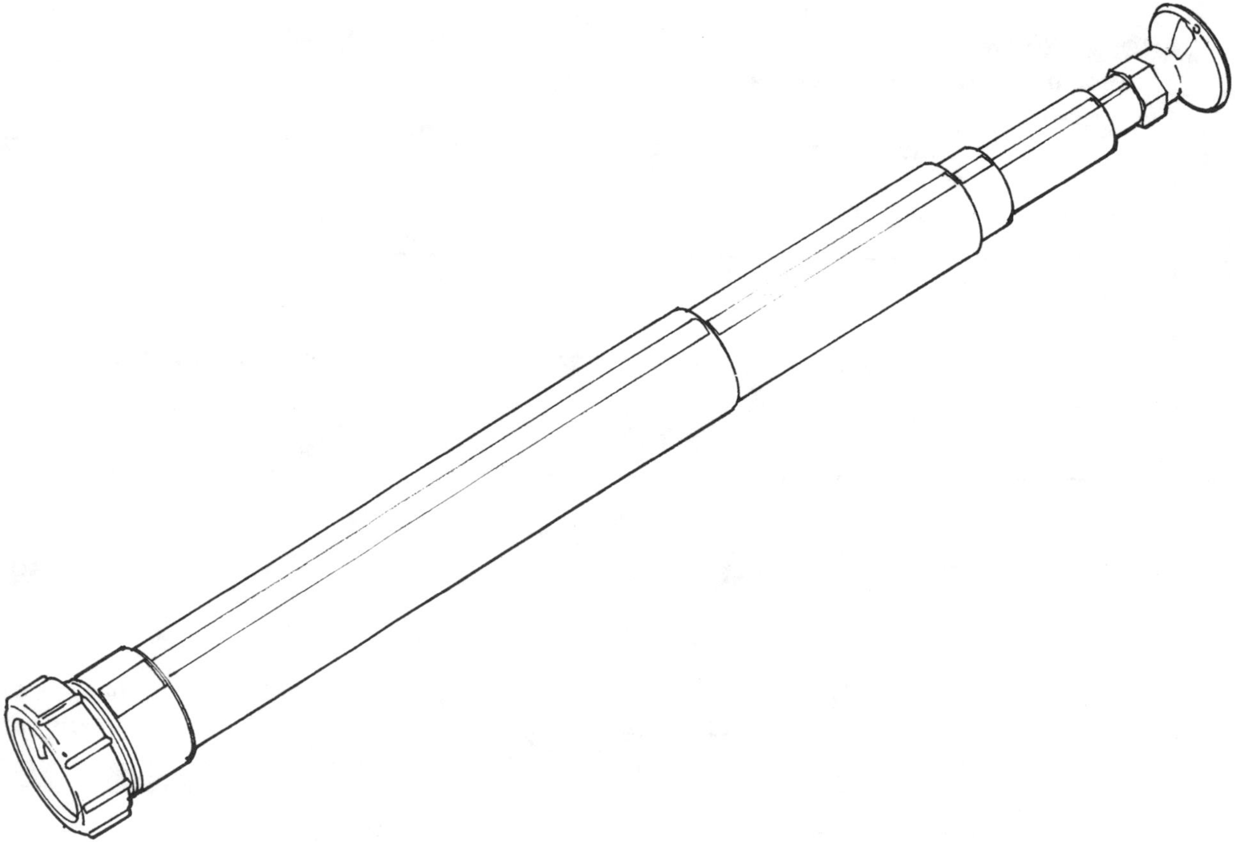

If it is daylight, take it outside and observe nearby objects. If it is night, try to observe the craters of the Moon, or try to locate Jupiter and its moons! DO NOT LOOK AT THE SUN WITH YOUR TELESCOPE BECAUSE THE SUN'S RAYS COULD SERIOUSLY INJURE YOUR EYES.

Galileo died in 1642. On Christmas Day of the same year, the great English scientist, Isaac Newton was born. Throughout his childhood, Newton spent most of his time alone, reading and making mechanical toys. It is said that he made windmills, a sundial and a waterclock that really kept time. Newton also enjoyed flying kites. Legend has it that he once attached a lantern to a kite he was flying at night. It greatly alarmed the people in the town who thought it was a comet!

When Newton entered Trinity College at Cambridge University, many people still believed that the sun orbited around the Earth. During his three years at the university, he took courses in mathematics, Greek and Latin. He didn't enjoy most of his studies and he didn't do very well in them. It was his tutor in physics and optics, Dr. Isaac Barrow, who opened up a whole new world for Newton. He introduced him to telescopes and the theories of light. Newton began grinding his own lenses and building his own telescopes. While observing the stars, Newton also began to develop his own theory of light.

Even though he ground his own lenses very carefully, Newton still had trouble with rainbows of color around the images he observed. Other astronomers had the same problem. They found that the wider the lens, the worse the rainbow distortion became. For this reason, there were no really big telescopes. After experimenting with the problem, Newton soon realized that the rainbows were not caused by the lenses, but by light itself.

Newton discussed this problem with Professor Barrow, who suggested that he read Kepler's book on optics. Even though he found more questions than answers in the book, it did point the way for him to make discoveries that changed both the science of optics and the science of astronomy.

Sir Isaac Newton (1642-1727) *Culver Pictures*

Newton began experimenting with single **prisms** and combinations of prisms to try to discover how the light was refracted when it passed through them. He set up window screens with a narrow slit or two that would allow only a little light to pass through. Then he measured both the angle at which the colors were bent and the images that were cast on the reflecting screens.

PRISM

A solid object whose bases or ends are the same size and shape and are parallel to one another. Each of the prism's sides has two parallel edges, like a six-sided pencil before it is sharpened. Transparent prisms are often made of glass with three-sided ends that can refract or separate light into the colors of the spectrum.

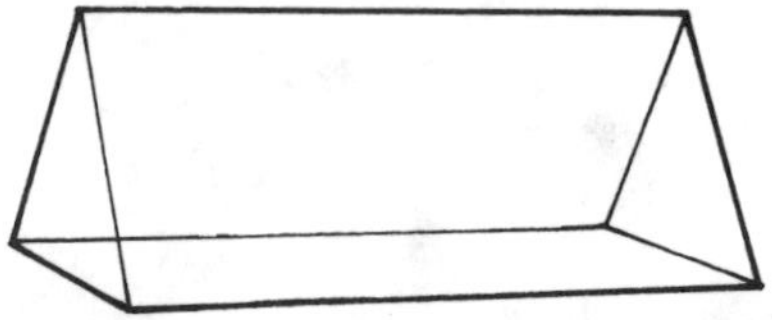

Three-sided Prism

Four-sided Prism

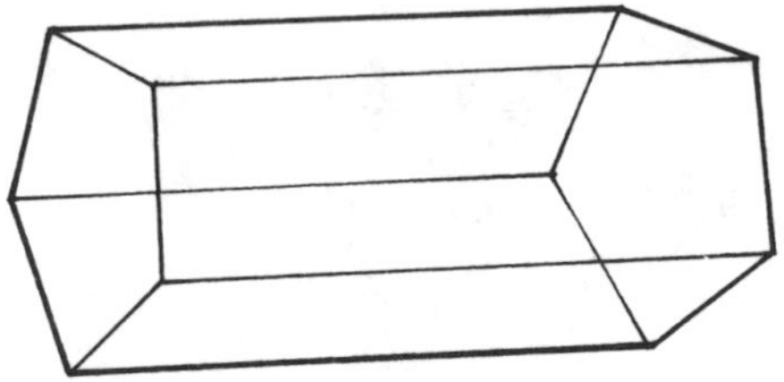

Five-sided Prism

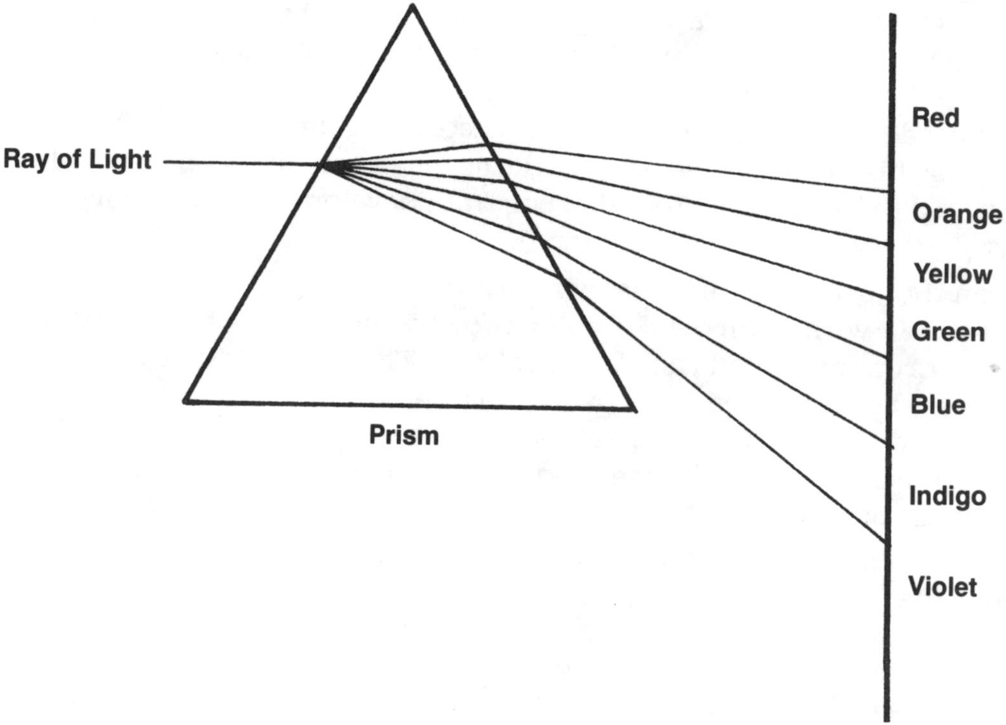

Newton soon realized that pure sunlight contains a whole rainbow of colors. With a prism he could divide a ray of sunshine into the colors of the **spectrum** and capture them on a screen. Newton discovered that different colors in the rays are refracted at different angles according to their wavelengths. Violet is refracted the most, while red is refracted the least.

SPECTRUM

The band of colors (red, orange, yellow, green, blue, indigo and violet) that is formed when light passes through a prism. A rainbow has all of the colors of the spectrum.

Newton also discovered that if he placed a second prism in the path of the light after it passed through the first prism and turned it so that all the colored rays entered one side of it, then he was able to control the light so that when it had passed through both prisms, it was pure white.

You can discover as Newton did, how light, which seems to be lacking any color, is in fact made up of all of the colors of the spectrum. Simply place a prism in front of a brightly lit window. The light will divide itself into the colors of the spectrum.

If you do not have a prism, you can still create a spectrum by using a glass of water instead. Place the glass slightly over the edge of a window ledge in bright sunlight. Then put a piece of white paper on the floor where the rays of sunlight are projected through the glass. A spectrum, or rainbow band, should appear on the piece of paper.

CHROMATIC ABERRATION

The color distortion in an image produced by a lens because different wavelengths of light are refracted differently.

Newton also discovered that one part of the light passing through a telescope lens could be refracted a certain degree while another part was refracted differently. This meant that one single lens would not be able to bring all the colors in the sunlight together at one point. This situation is known as **chromatic aberration**.

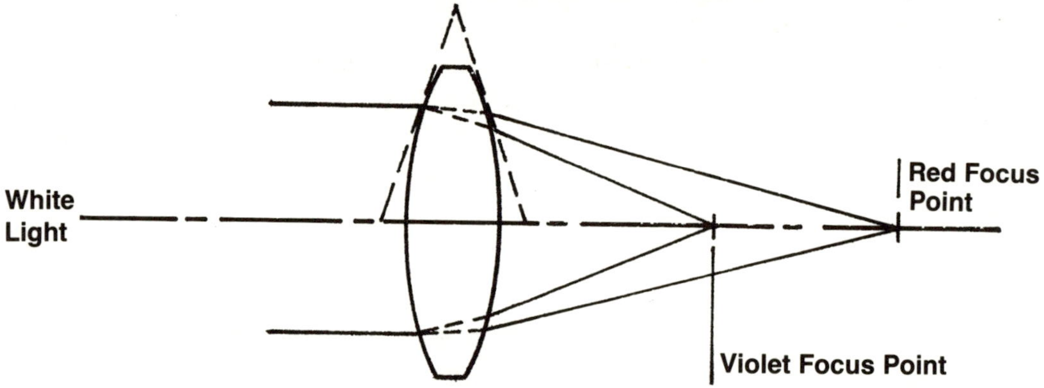

Chromatic aberration showing the different focus points for violet and red light. Notice how the lens section resembles a prism.

Today we correct this situation in cameras and telescopes by using an achromatic lens, which is a combination of lenses that produce images free of chromatic distortion.

By 1669, Newton was back at Trinity College as a professor. He was always considered a very absentminded professor, sometimes even forgetting to eat when he was working on an experiment. But he took his duties as a teacher seriously. Since he was supposed to deliver a certain number of lectures he did so with enthusiasm, even when no students showed up for the class!

In that same year, Newton tried to solve the problem of chromatic distortion appearing in his telescope by building a reflecting telescope. He used a curved mirror in this telescope, instead of a lens, to focus the sunlight, or the light from the stars and planets more precisely.

You can see for yourself how a curved mirror will focus light. Place a small magnifying mirror (like the type used when putting on makeup) in front of a light bulb. Use the mirror to focus the image of the light bulb onto a piece of white paper. When the light hits the reflective surface of the curved mirror, it is focused as it would be by a lens. By using a mirror in this way in place of a lens, Newton was able to overcome the problems caused by chromatic distortion.

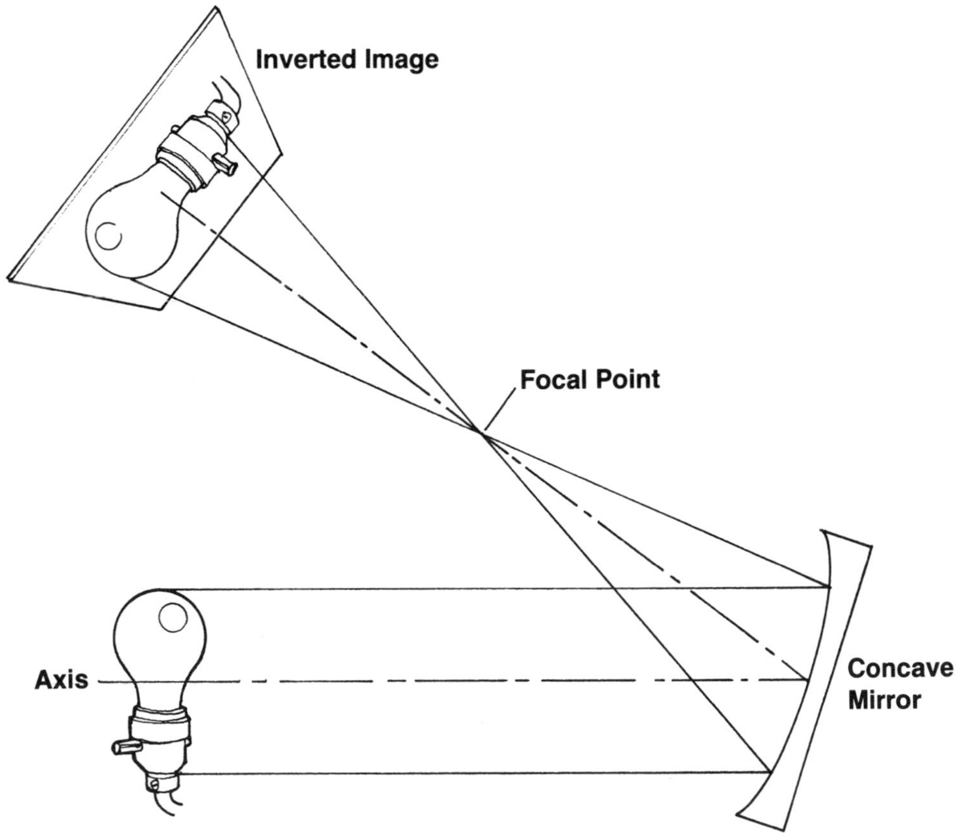

The front of Newton's telescope was left open. A concave mirror was placed at the back of the tube to collect the light. This object mirror reflected the image onto a second small, flat mirror which was placed inside the tube near the front. The secondary mirror was mounted at a 45 degree angle so that it reflected the image through a hole in the side of the tube. The only lens in the entire telescope, a double convex eyepiece, magnified the image which appeared without any of the rainbow colors seen in earlier instruments. The telescope itself was mounted on a movable sphere, so that it could be turned to any direction.

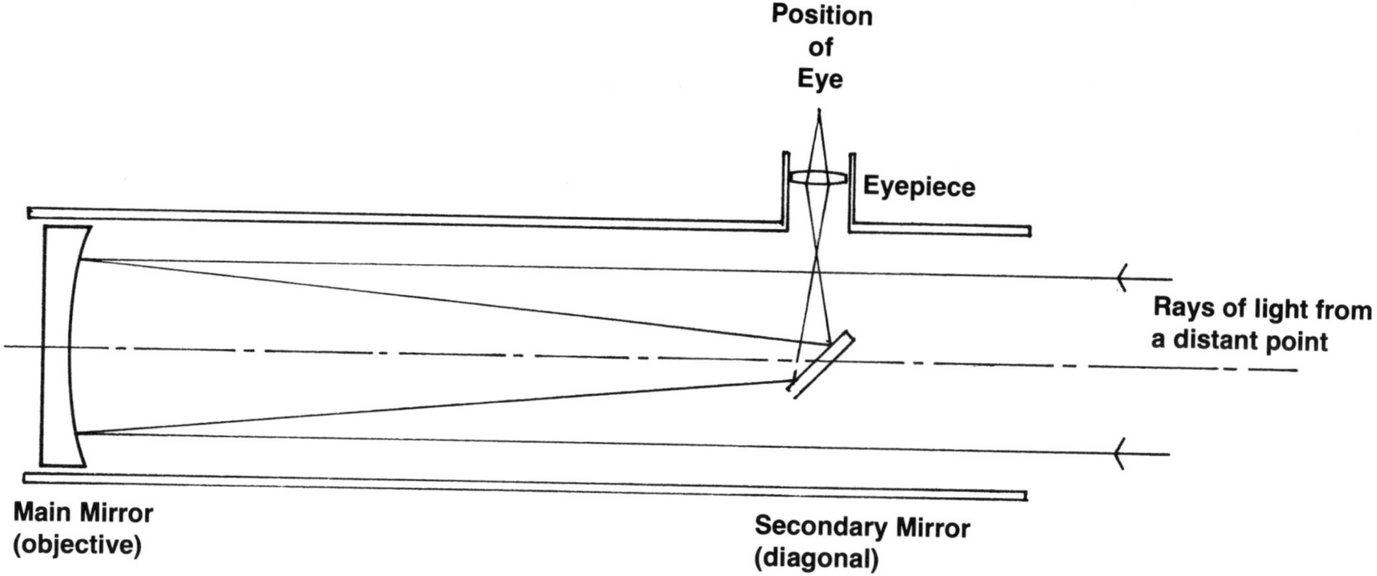

Diagram of Newton's reflecting telescope

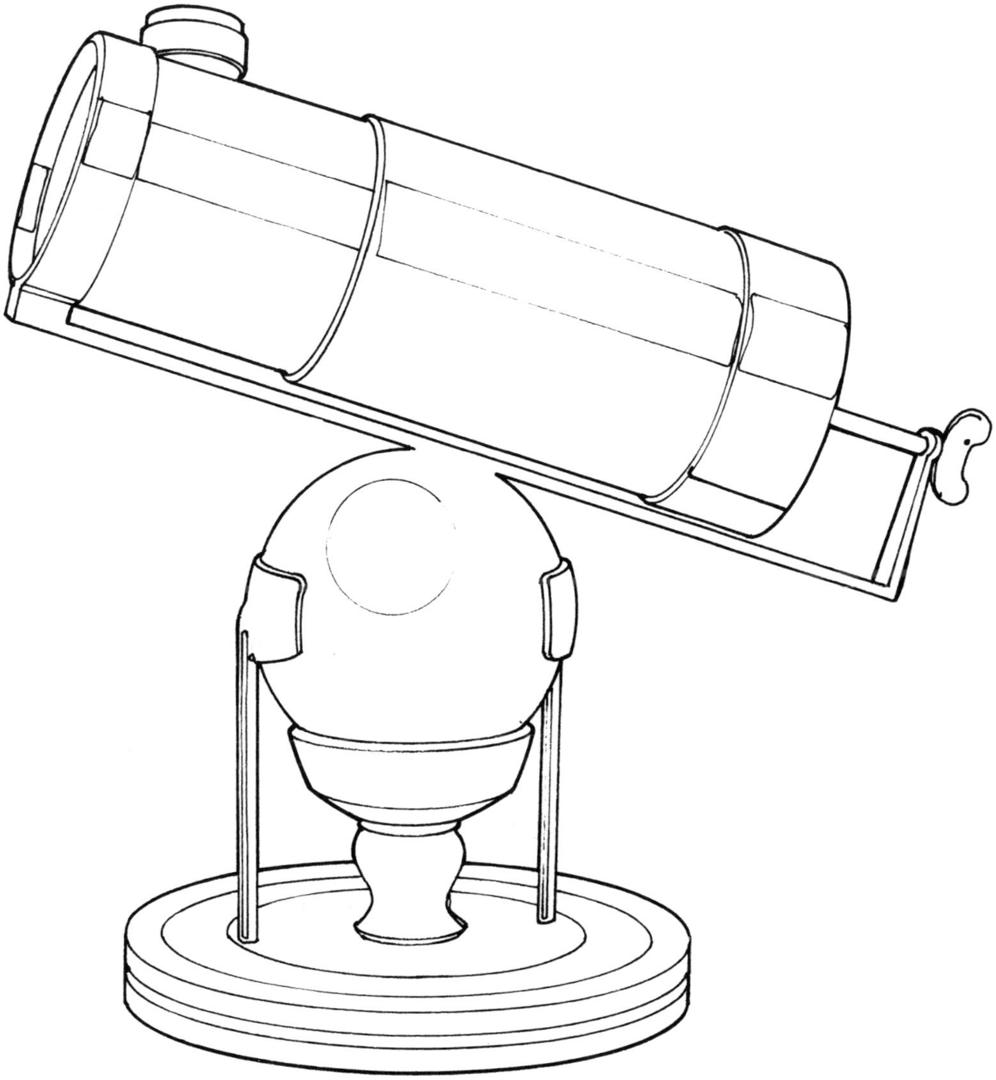

Newton's Reflecting Telescope

You can build an inexpensive, easy-to-construct, 3-inch Newtonian reflecting telescope using lenses and standard PVC plumbing parts. In order to do so, you will need the following materials:

Hand drill and 5/32 bit
Countersink
X-Acto knife
Scissors
File
PVC cement
White glue
Pliobond glue or double-faced tape
Tape
2-ply bristol board for rolling tubes
Heavy and fine sandpaper
Flat-black spray paint
Three 8•32 x ¾″-flathead machine screws; three 8•32 nuts; and three 8•32 star lock washers
Three plumbing faucet washers, ⅝″ diameter, ¼″ thick (They have to fit the 8•32 machine screw.)
One 32″ section of a 4″ PVC pipe

One 2″ slip by thread bushing—½″ threaded center

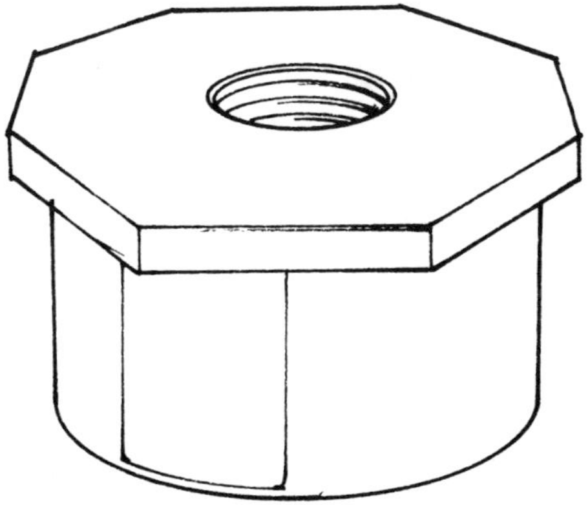

One PVC slip by barb adaptor

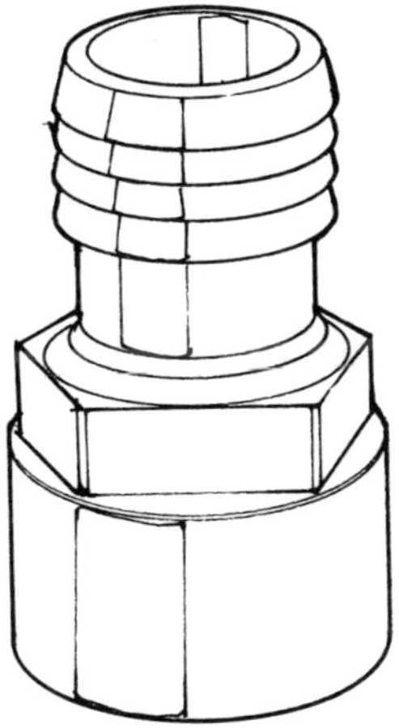

The other PVC pieces you will need to build the equatorial mount are described on pages 151-153.

You will need to order the following materials from an optical supply house, or from Edmund Scientific:

One 3″, ¼ wave mirror with a 30″ focal length
One 3″ telescope mirror mount
One 21mm x 30mm diagonal mirror holder and mirror
Two 21mm x 40mm plano convex lenses

MAIN TUBE

The basis of this telescope is a 32″ section of 4″ PVC pipe. This pipe measures 4″ inside and 4½″ outside. The ends of the pipe must be filed true. In order to do this, take a piece of paper 5″ x 15″ and wrap it tightly around the pipe. Make sure that the top surface is continuous. Slide this paper tube to the lowest point on the end that is true, or even. Inscribe a line around the tube at this point. Save this paper tube for subsequent markings of dimensions.

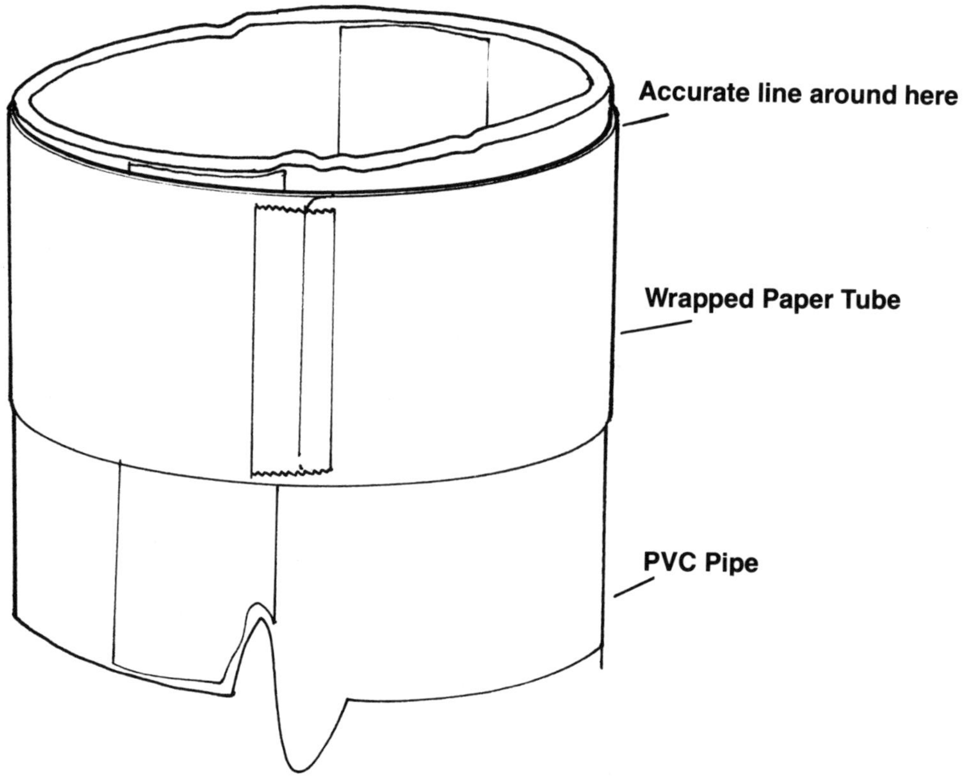

Use a file to true up the end of the pipe, filing accurately to the inscribed line. Repeat this process at the other end of the pipe, trueing at exactly 32″.

Lay the trued pipe on a perfectly flat surface. Inscribe a line AB the length of the pipe. An easy way to guarantee a line parallel to the sides is to prop the pen on some flat object and slide the pen and object over the table, marking the pipe as shown below.

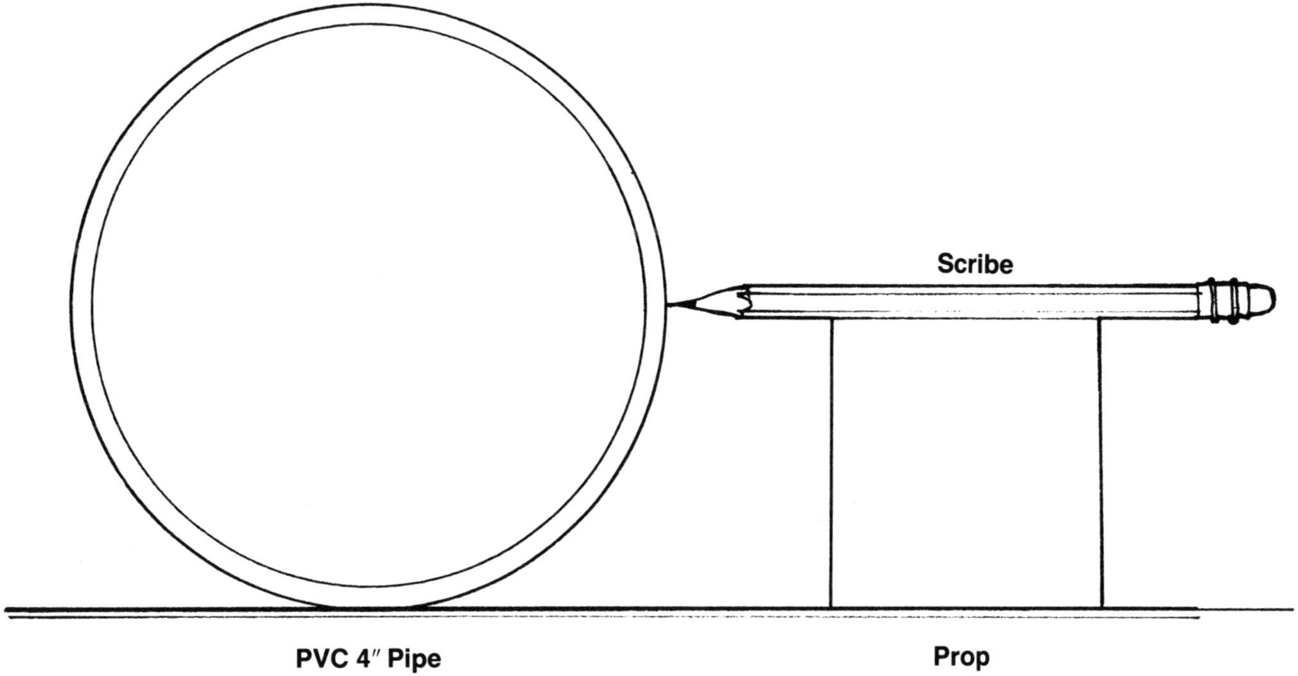

TEMPLATE

A pattern, or gauge, used as a guide in making something accurately.

Along this line, lay out the dimensions from the following drawing. Use the paper tube **template** from step #1. Slide it down the pipe to each dimension and mark lines around the pipe.

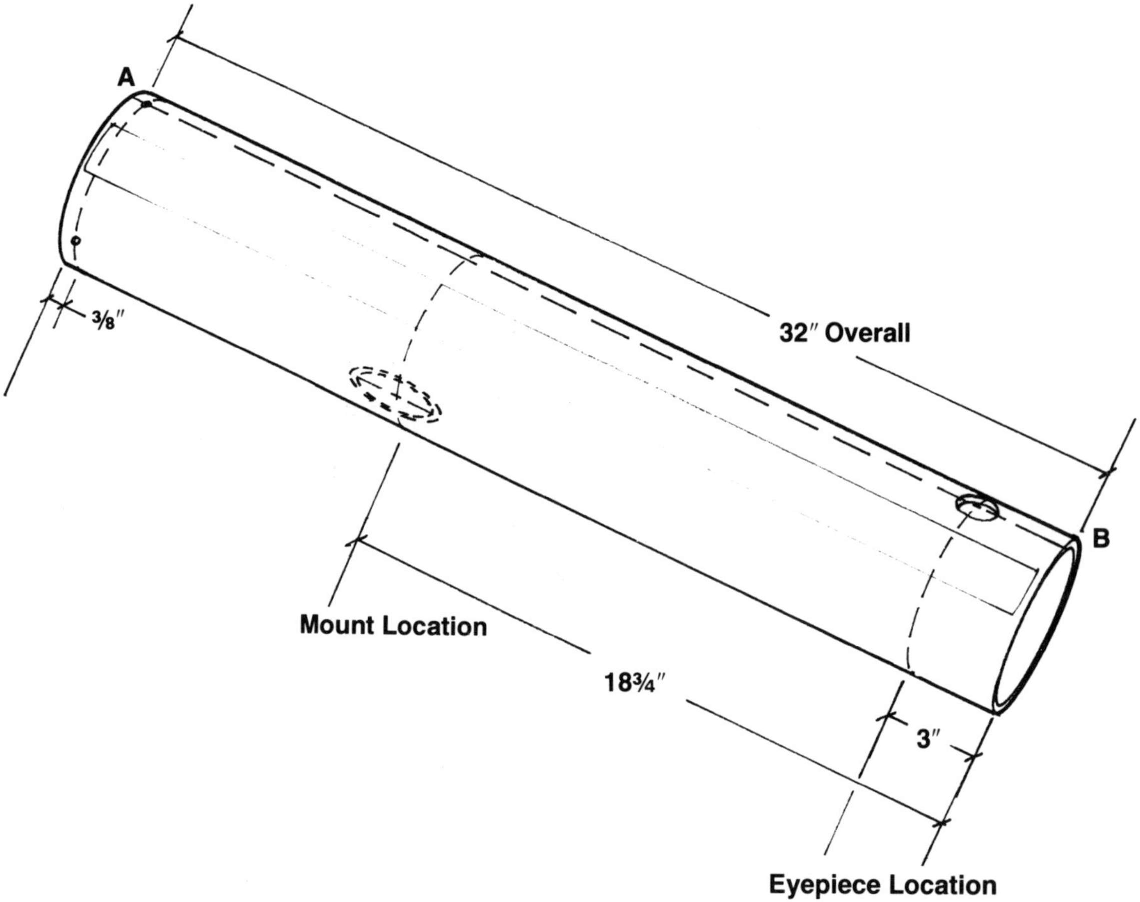

Lay out and drill three 5/32″ holes, 120 degrees apart, along the 3/8″ line at the mirror end of the pipe. Make sure that one of the three holes is on the length line *AB*. Countersink these holes until an 8·32, flathead machine screw fits flush.

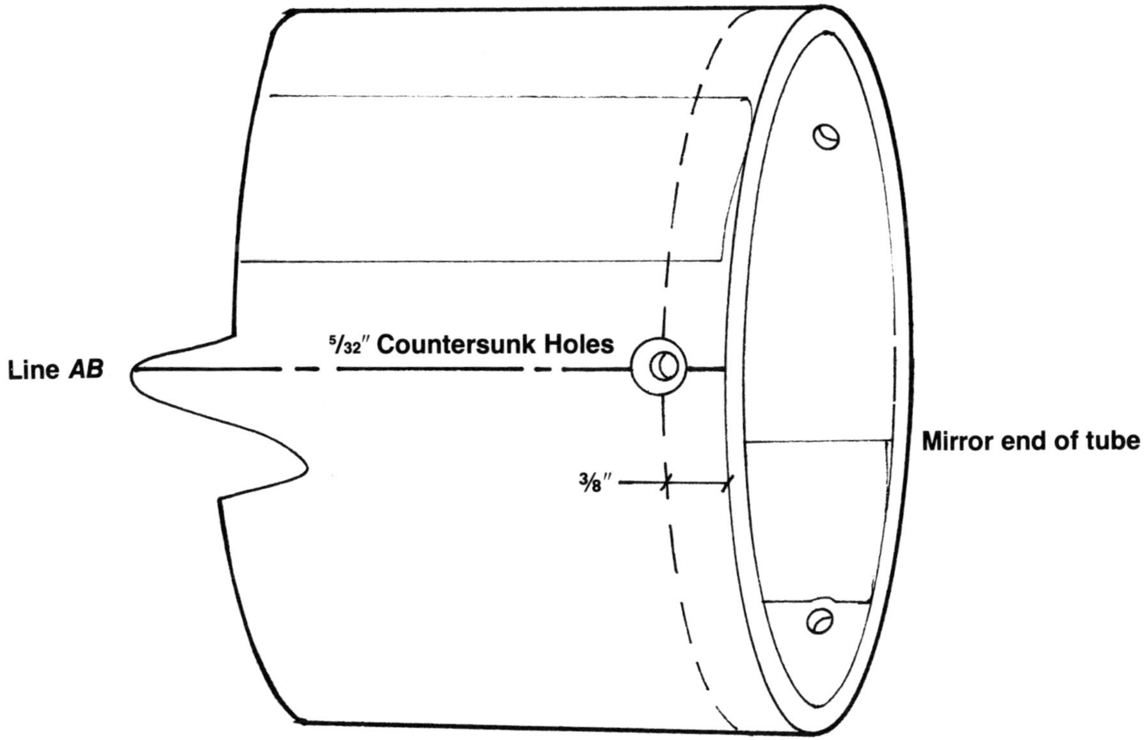

Make a 1⅛″ hole for the eyepiece at the 3″ line at the eyepiece end of the pipe.

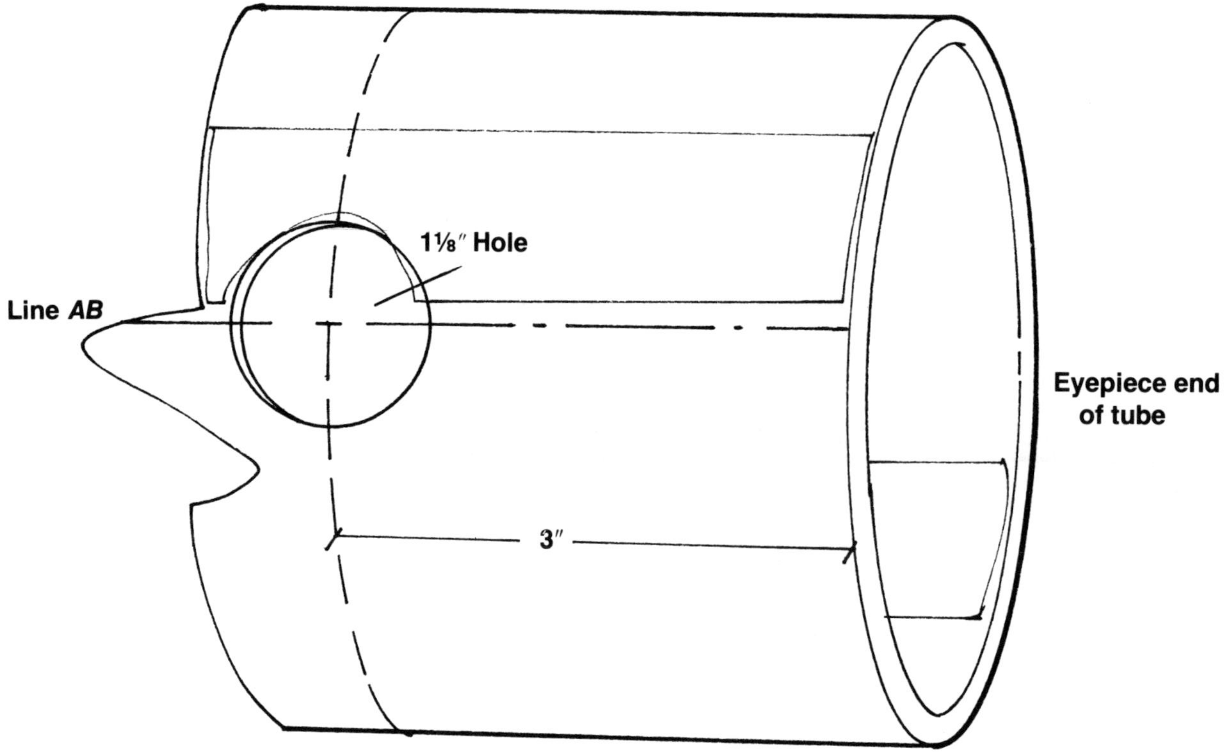

Sand off the rough edges and spray the inside of the pipe flat black.

Tape a sheet of sandpaper around the curve of the pipe, rough-side up. Sand the eyepiece holder and mounting lug by moving them lengthwise over the sandpaper until they fit the contour of the pipe perfectly.

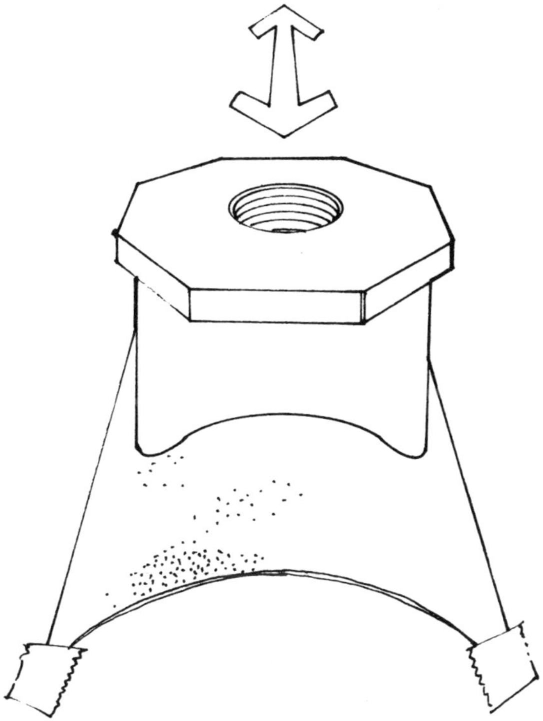

Bottom of the mounting lug unit is sanded to match the curvature of the main tube.

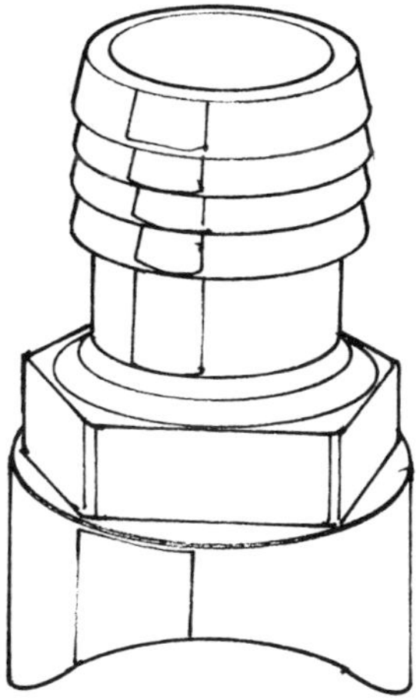

Eyepiece-holder base sanded to fit diameter of main tube.

Glue the eyepiece holder in place with plumber's PVC glue or cement. Center it over the 1⅛″ hole. Glue the mounting lug in place along the line, 135 degrees from the line *AB*. Set the unit aside to dry.

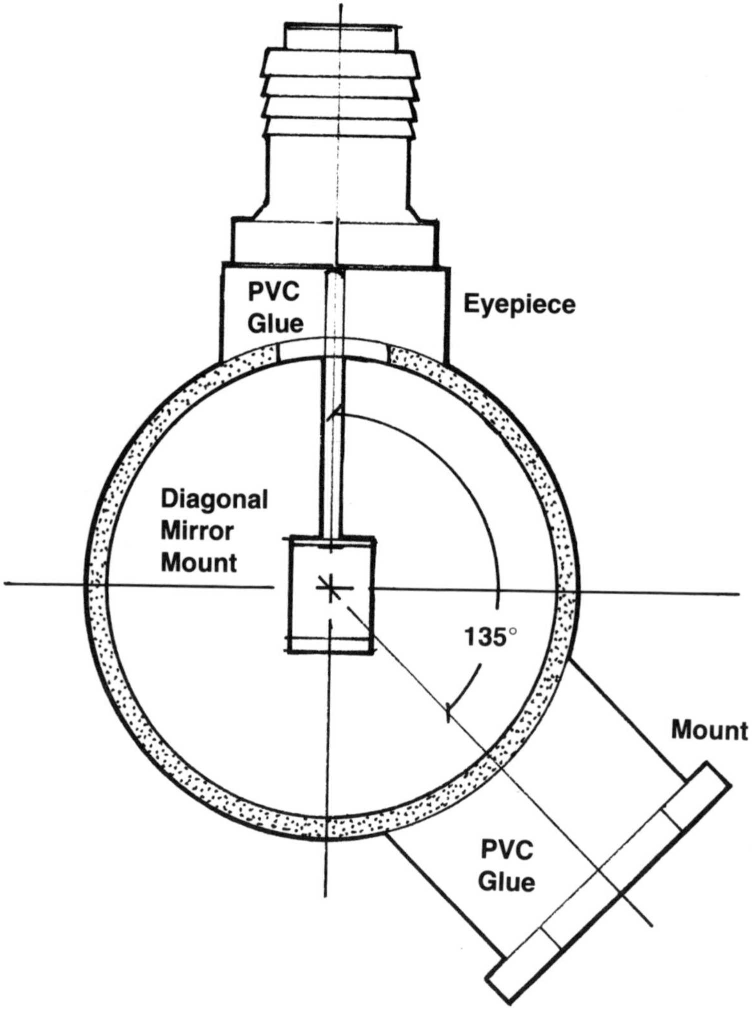

MIRROR SYSTEM

The 3-inch diameter, ¼ wave, 30″ focal length, objective mirror is purchased separate from its mount. The mirror mount must be assembled according to its instruction booklet and then the main objective mirror is mounted to it.

The thumbscrew/spring system allows for total adjustment of the mirror once it is assembled and mounted to the main telescope tube. You will need three 8•32 x ¾″-flathead machine screws, nuts and star lock washers. You will also need three ¼″ thick spacers, drilled to 5/32″. Plumbing faucet washers work perfectly.

Mount the fully assembled mirror and mount unit to the mirror end of the telescope tube as shown.

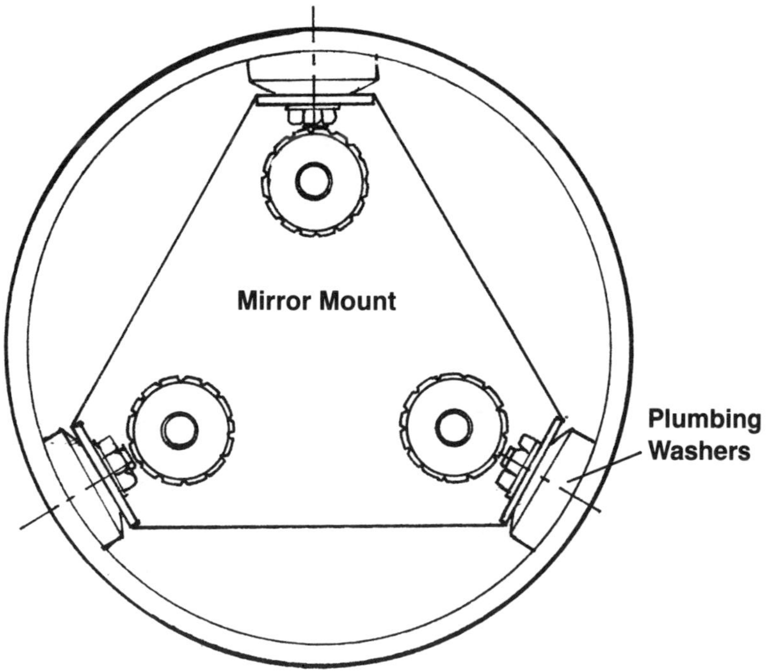

The 21mm x 30mm diagonal mirror is also purchased separate from its mount. Adhere the mirror to the mount using pliobond glue, or double faced tape.

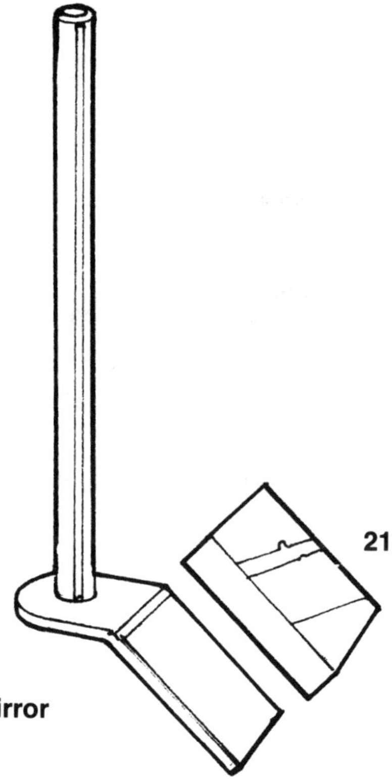

Drill a $^{5}/_{32}''$ hole through the main tube, along line AB, directly behind the eyepiece mount.

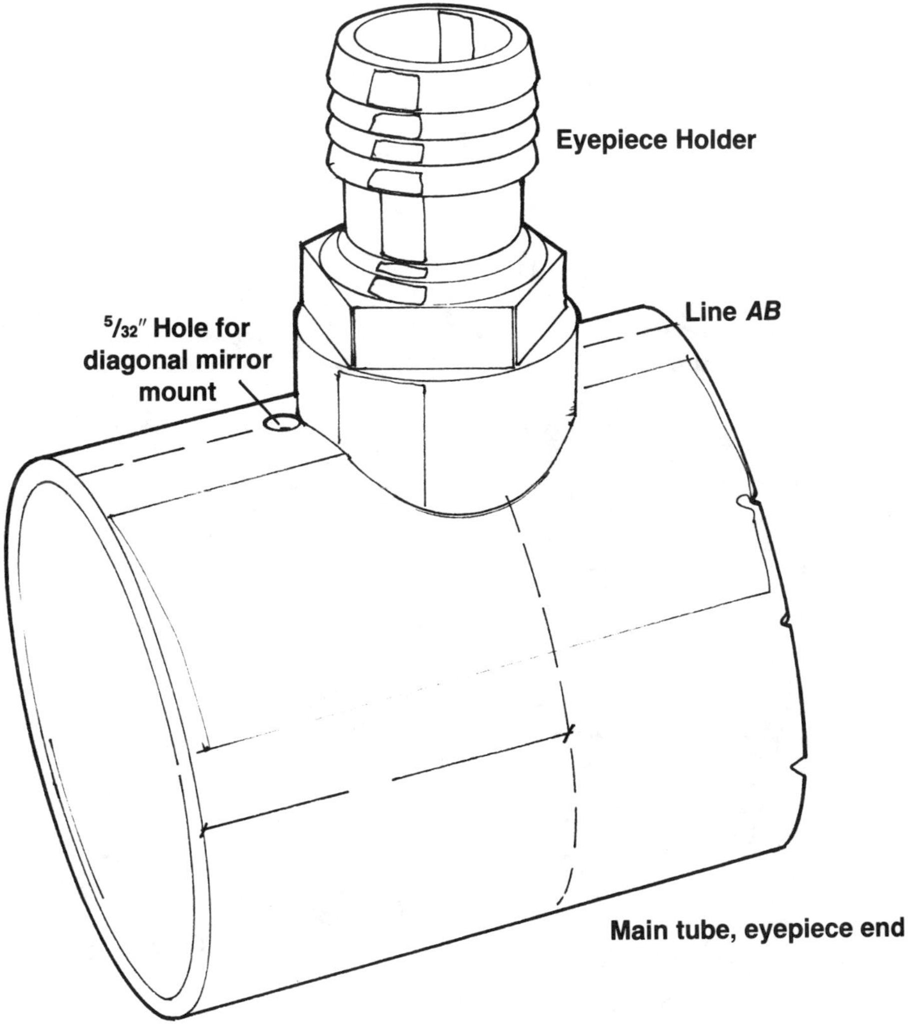

Insert the mirror and mount unit through the hole from the inside and position as shown below.

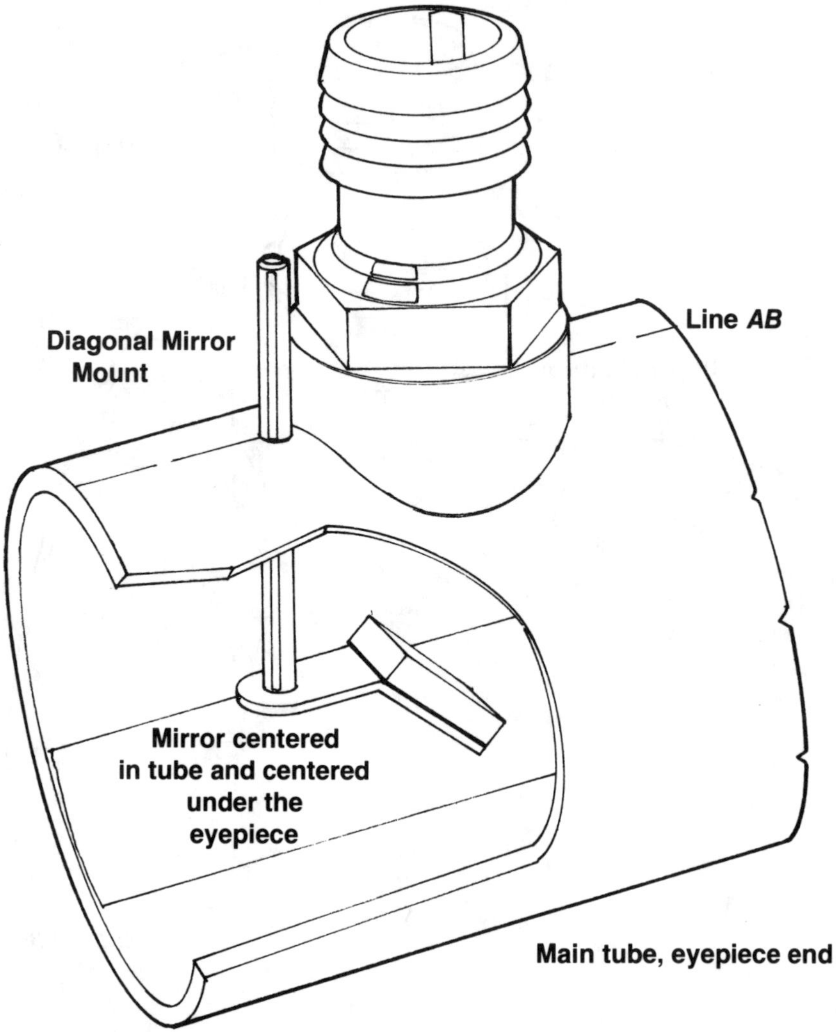

EYEPIECE

To build your own **Ramsden eyepiece** you will need a 13/16″ mandrel to wrap the spacer tube over. Wrap and glue enough heavy paper to have a 21mm x 3″ tube.

Wrap and glue a second 3″ tube (the lens tube) over the first, being careful not to accidentally glue the 2 pieces together. Continue wrapping the lens tube until it slides into the eyepiece holder on the main tube.

Slide the tubes apart, trim the lens tube to 2⅞″. Cut the spacer tube into three units: ¼″, 15/16″, and 1¼″. Spray the inside of the spacers flat black.

Make the light stop out of cardboard or paper, or find a plumbing washer that will do the job. Lay out the tubes, spacers, light stop and lenses as shown below.

RAMSDEN EYEPIECE

The Ramsden eyepiece was invented in 1783. It reduced the spherical aberration problem found in earlier eyepieces. Ramsden eyepieces are particularly suited for use with reflecting telescopes that have short focal lengths.

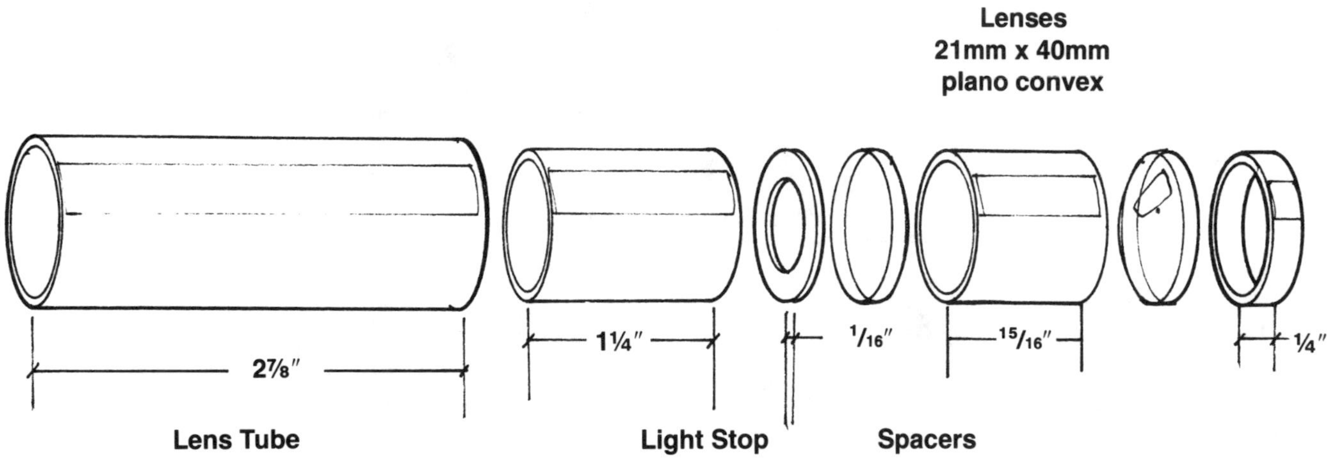

Diagram of Ramsden eyepiece

Glue the ¼″ spacer into one end of the lens tube. This will be the only spacer glued in order to be able to remove and clean the lenses if necessary. Insert 1 of the lenses into the tube, so that the flat side is against the ¼″ spacer. Follow with the 15/16″ spacer, and the second lens, curved side inward. Add the light stop followed by the 1¼″ spacer. The eyepiece is now finished and can be positioned in the holder as shown.

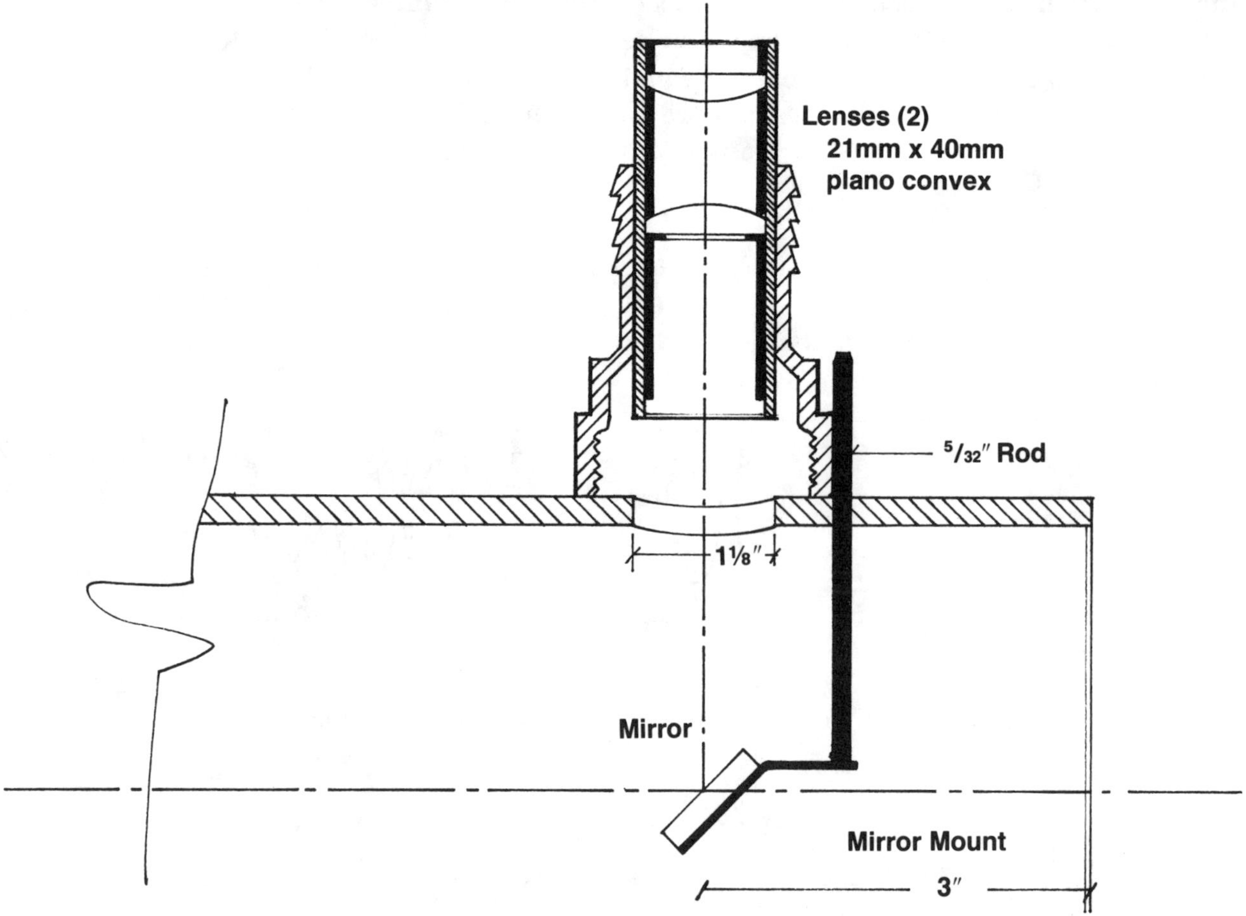

COLLIMATION

Before you use your telescope, you must adjust and align the mirrors. First remove the eyepiece and look through the eyepiece holder. Aim the telescope toward a brightly lit ceiling. Adjust the diagonal unit until it is centered under the eyepiece holder opening. The round objective mirror should be centered in the diagonal mirror.

COLLIMATION

The adjustment of the line of sight of the telescope.

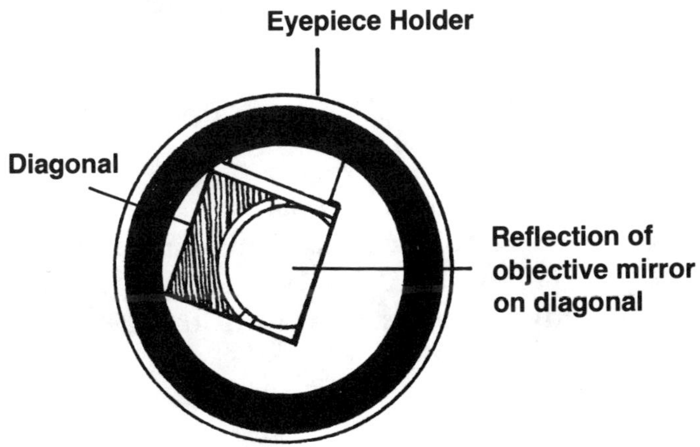

Diagonal off-center

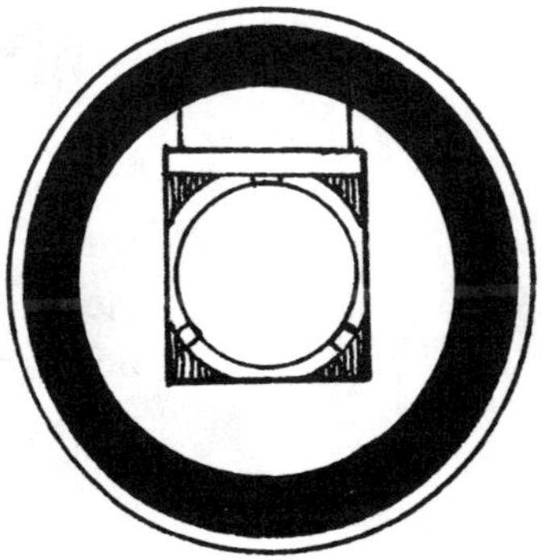

Diagonal aligned

You will be able to make this adjustment by using gentle pressure to twist the diagonal in its socket. Alternate looking through the eyepiece holder and looking down the eyepiece end from a few feet away. Make sure that the diagonal unit is also centered in the main telescope tube.

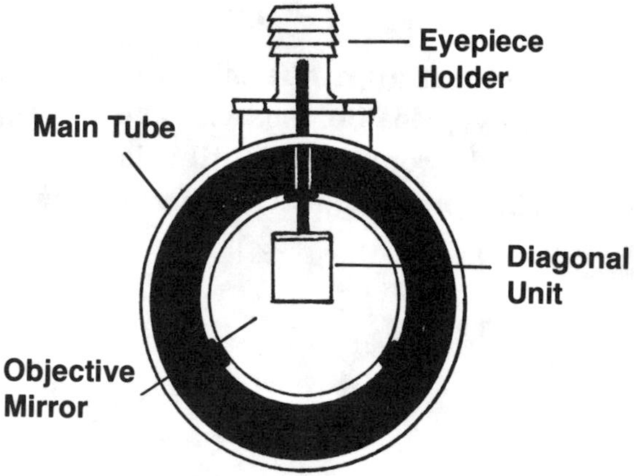

Diagonal too high

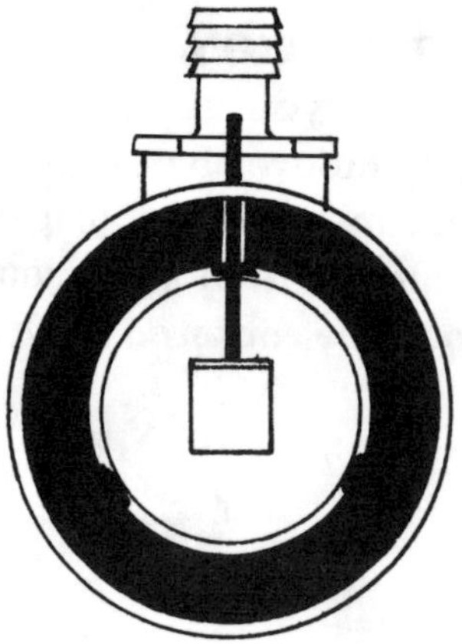

Diagonal centered

Once the diagonal is aligned, you can proceed to align the objective mirror. Look through the eyepiece holder again. This time, note the position of the reflection of the diagonal in the objective mirror. Turn the 3 adjustment nuts on the objective mirror mount until the diagonal unit's reflection is centered.

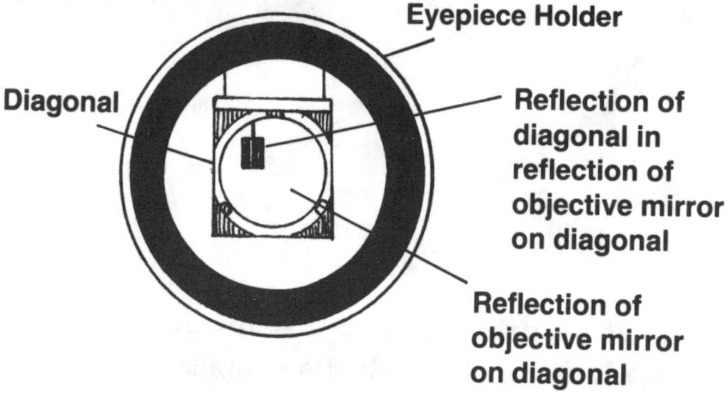

Objective mirror not aligned

Proper alignment

EQUATORIAL MOUNT

The mount for the telescope requires the following standard PVC pipe-fitting components:

One 1″ outside slip bushing and one ¾″ inside slip bushing

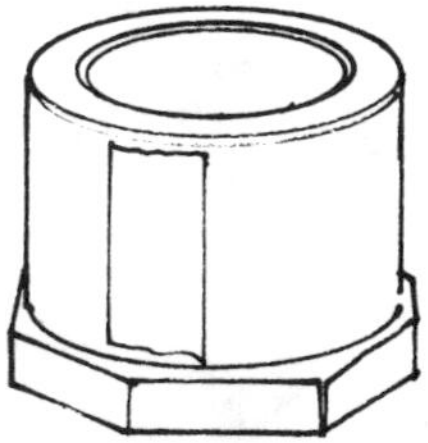

One 2″ slip by thread bushing with ½″ threaded center with 4 webs (for counterweight)

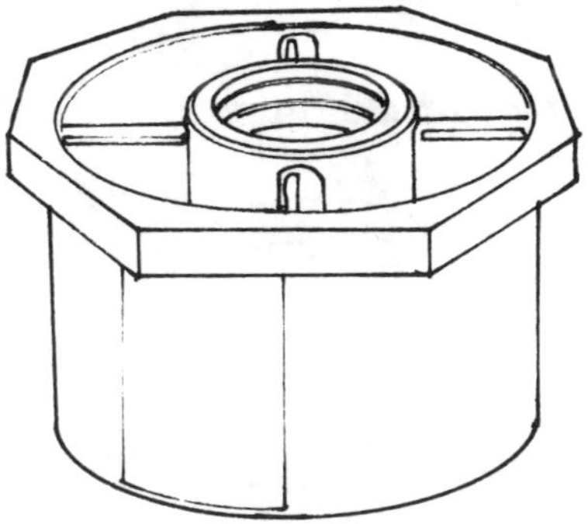

One ½″ x 4″ PVC straight pipe nipple and one ½″ x 12″ PVC straight pipe nipple

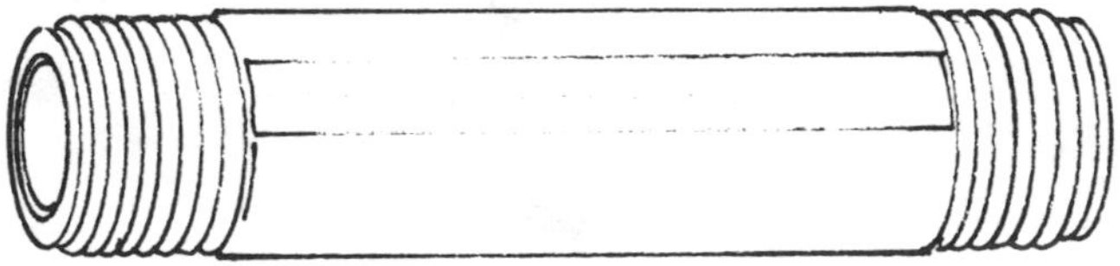

One ¾″ slip cap

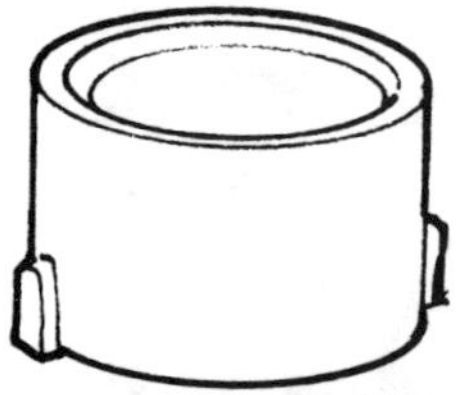

One 1″ slip tee

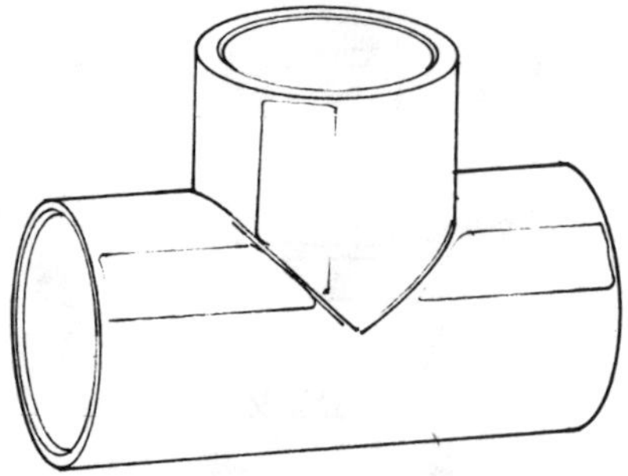

Two 1″ outside slip by thread bushings with ¾″ threaded inside diameters

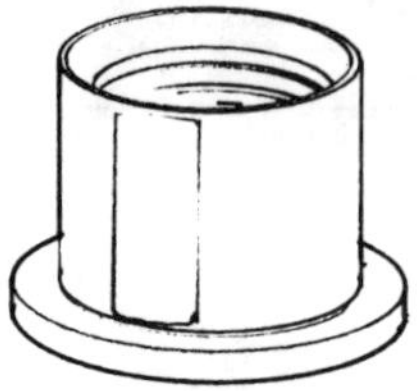

One ¾″ close nipple

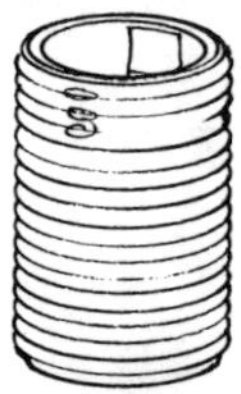

One 1″ 45-degree slip elbow

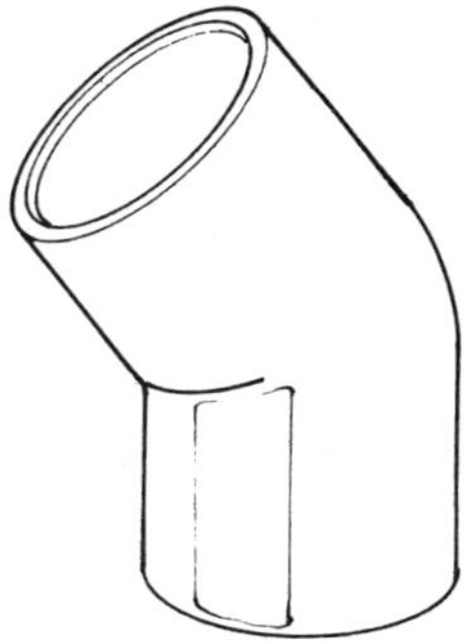

One 1″ outside slip bushing, ¾″ inside slip (bottom half), ½″ thread coupling

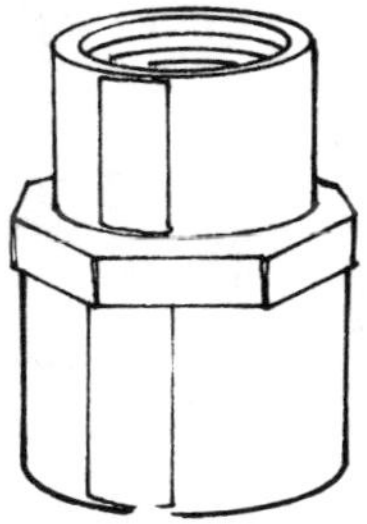

One ½″ threaded coupling

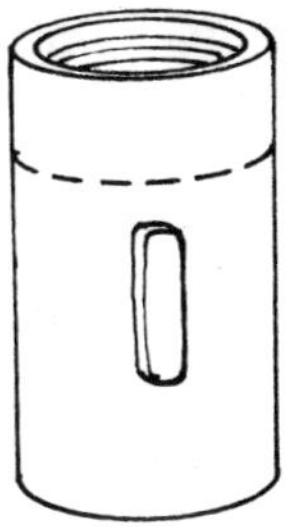

TAP

A tap is a tool used for cutting an internal screw thread.

Start the mount construction by filing the "ears" off of the ¾″ slip cap. This piece is normally fit over the end of a pipe, but when trimmed it fits inside one end of the 45-degree slip elbow. Find the center of the slip cap and drill and **tap** a ¼ x 20 threaded hole. This allows the unit to be bolted to a standard photography tripod. Turn the cap over and sand down the open end, checking the fit until the flat bottom fits flush when inserted into the elbow. Use PVC cement to glue the cap in place.

Drill and tap ¼ x 20 threaded hole.

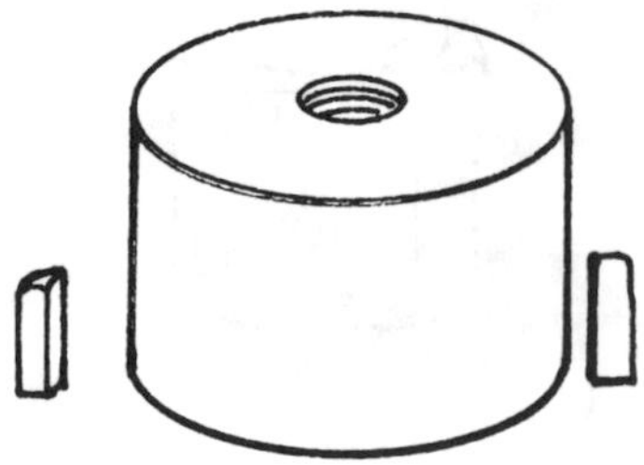

Remove ears. (Sometimes the piece comes without ears.)

Glue one of the 1″ slip by thread bushings into the other end of the elbow.

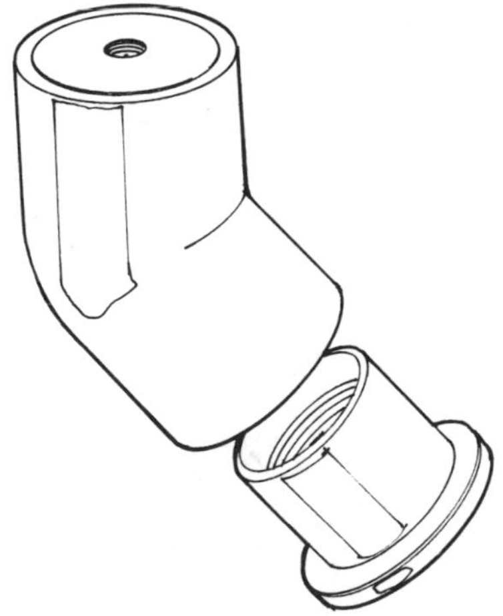

Glue the other slip by thread bushing into the bottom of the 1″ slip tee. Set these parts aside to dry.

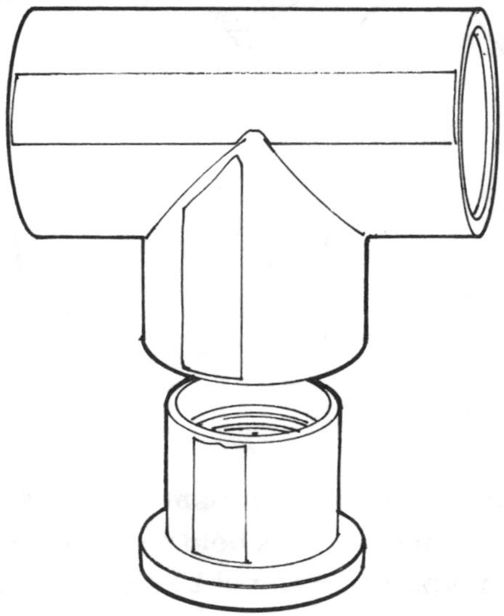

The tee and elbow units are joined together with a 2″ piece of ¾″ close nipple. This nipple is meant to be a pivot point which allows the tee section to pivot on the fixed elbow piece.

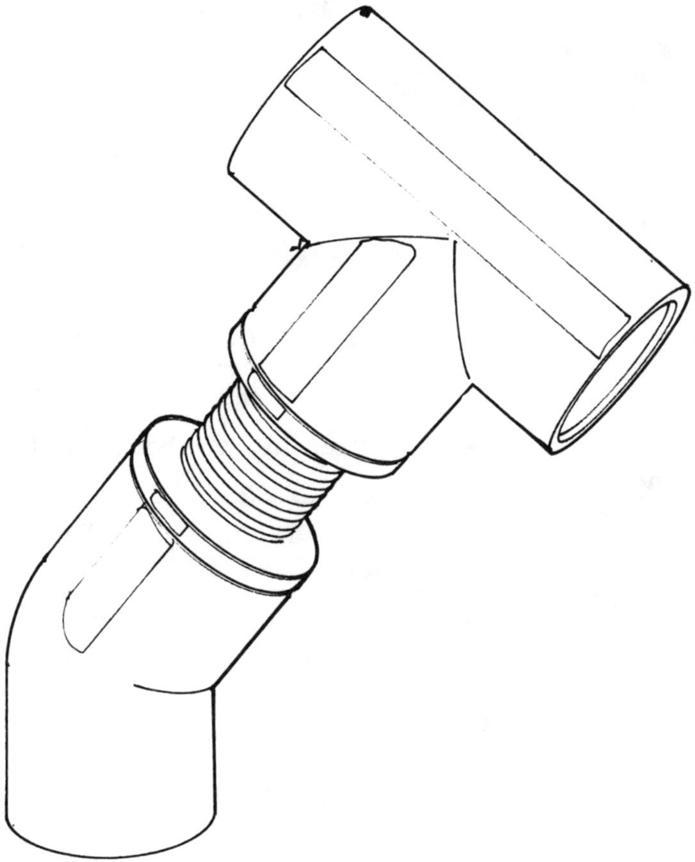

The counterbalance/pivot mechanism is made up of two outside slip bushings and inside slip bushings held together by a straight pipe nipple. The amount of free pivot is regulated by tightening the mechanism.

Two of the pieces need to be slightly modified as follows: The 1″ outside slip bushing and the ¾″ inside slip bushing must be reamed out or enlarged inside until the ¼″ x 4″ PVC straight pipe nipple rotates freely inside.

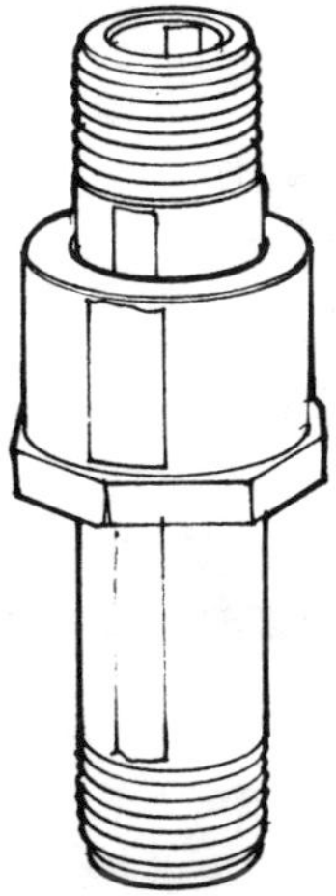

Cut off ¾″ from one end of the ½″ threaded coupling and cement it inside the large end of the 1″ outside slip bushing/¾″ inside slip bushing/½″ threaded coupling combination piece.

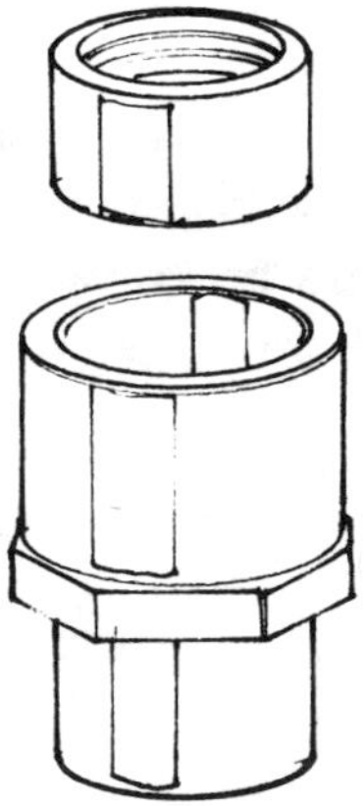

The final assembly is as follows:

1. Screw the ½″ by 4″ PVC straight pipe nipple into the mount on the telescope body.
2. Slide the bushing over the nipple, hexagonal end first.
3. Slide the 2″ tee over the bushing.
4. Thread the modified bushing onto the nipple.
5. Add the 12″ section of ½″ PVC pipe and the 2″ slip by thread bushing as shown on page 159.
6. You may want to increase the counterweight by gluing lead shot inside the webbed sections. Epoxy, or white glue, may be used to secure the shot in place. **REMEMBER NOT TO LOOK AT THE SUN THROUGH YOUR TELESCOPE BECAUSE IT WILL SERIOUSLY DAMAGE YOUR EYES.**

EQUATORIAL

Having a support with two perpendicular axes, one of which is parallel to the Earth's rotational axis.

The type of mounting you have constructed for your telescope is known as an **equatorial** mounting. It has two axes. When properly adjusted they allow you to follow the motion of a star or planet from east to west with very simple adjustments of the mounting. In order to do this, point the axis at the North Star (Polaris). Then point the axis attached to the base (parallel to the telescope tube) at the North Star.

Your telescope and its mounting should be attached to a stable tripod. Any vibration will cause the image you are observing in the eyepiece of your telescope to tremble. The greater the magnification of a telescope, the more serious a problem this becomes. A heavy-duty camera tripod will normally make a very good base for your telescope.

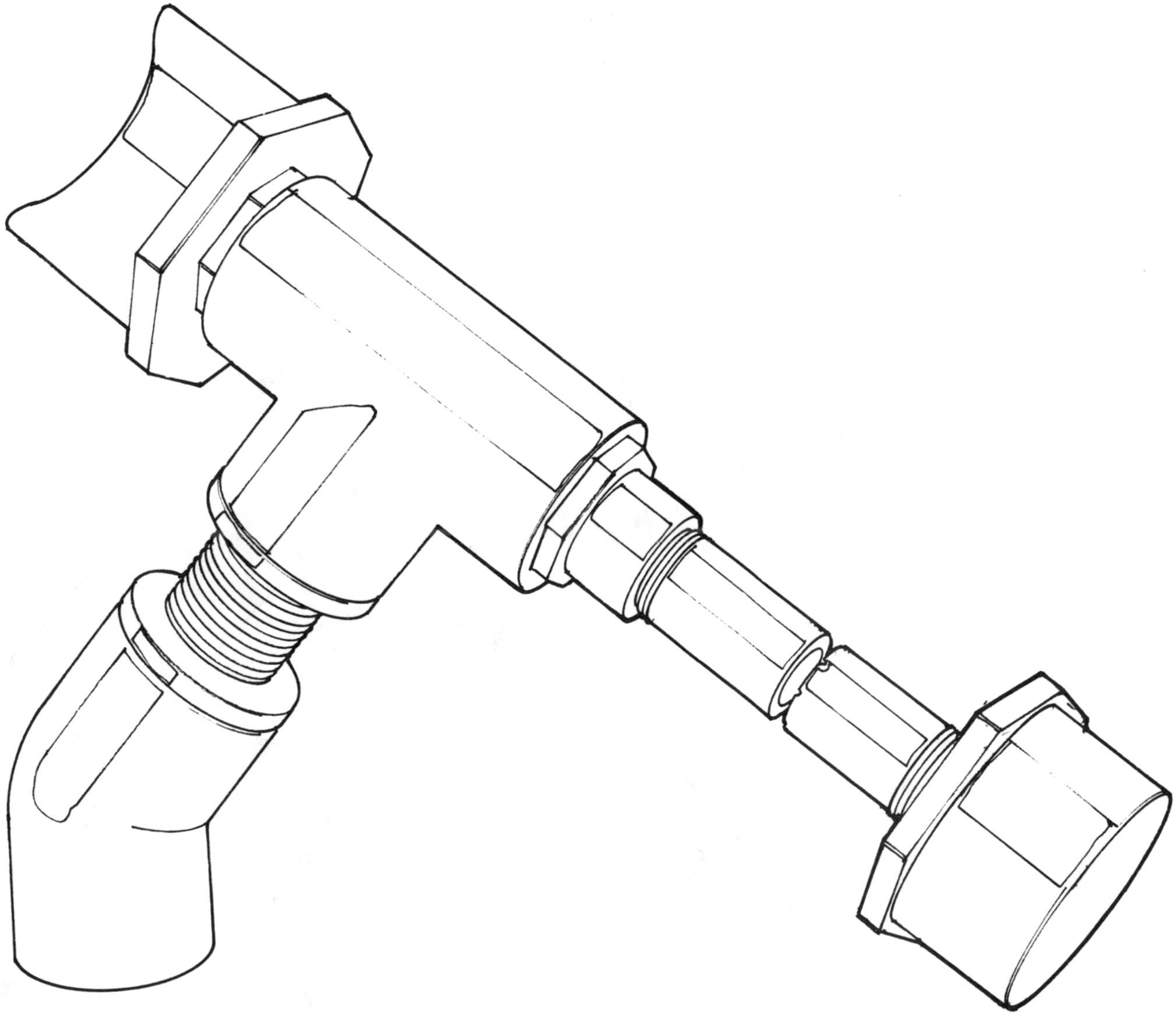

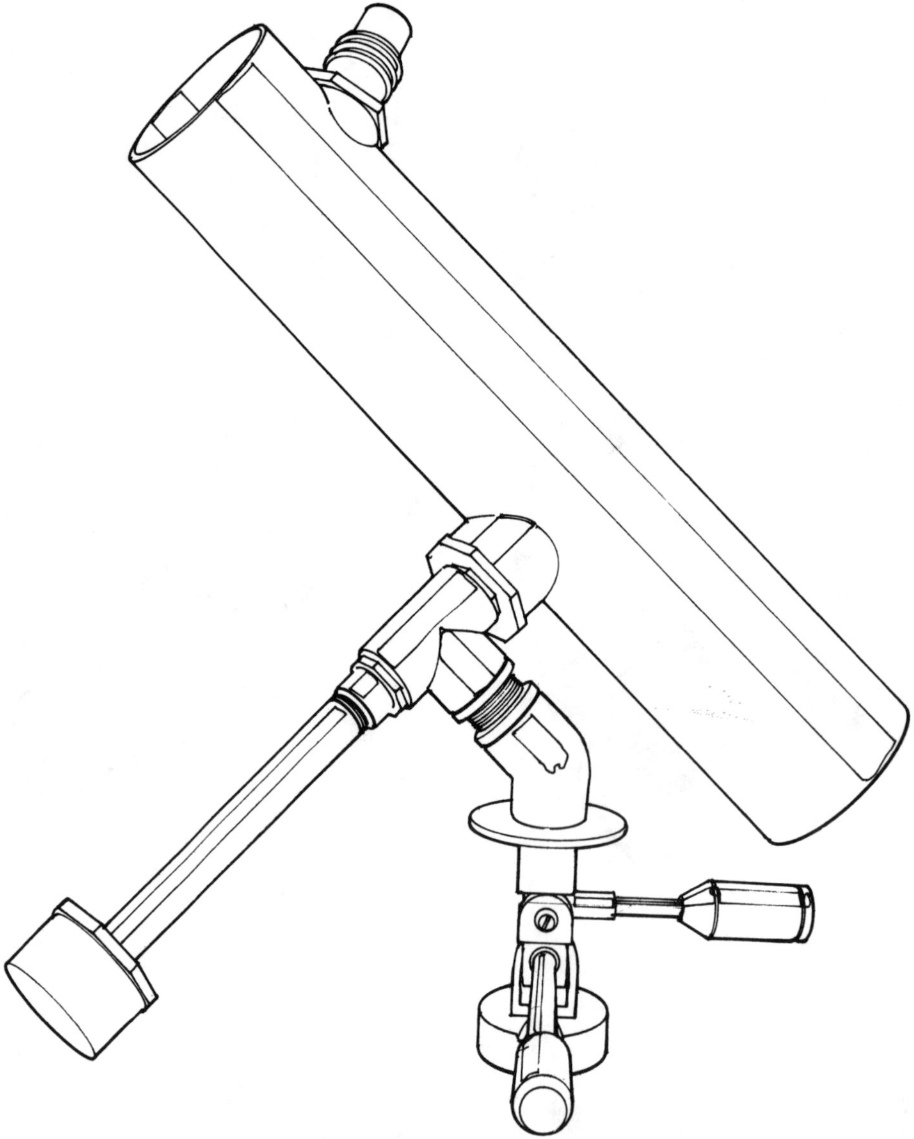

Reflecting telescope mounted on a camera tripod.

If you do not have a camera tripod that you can use to mount your telescope, you can attach the mount to a piece of ¾″ plank. Use a 1½″ x ¼•20 bolt through the bottom of the plank. With a C-clamp, clamp the board to the top of a stepladder. Mount the telescope onto the top of the bolt.

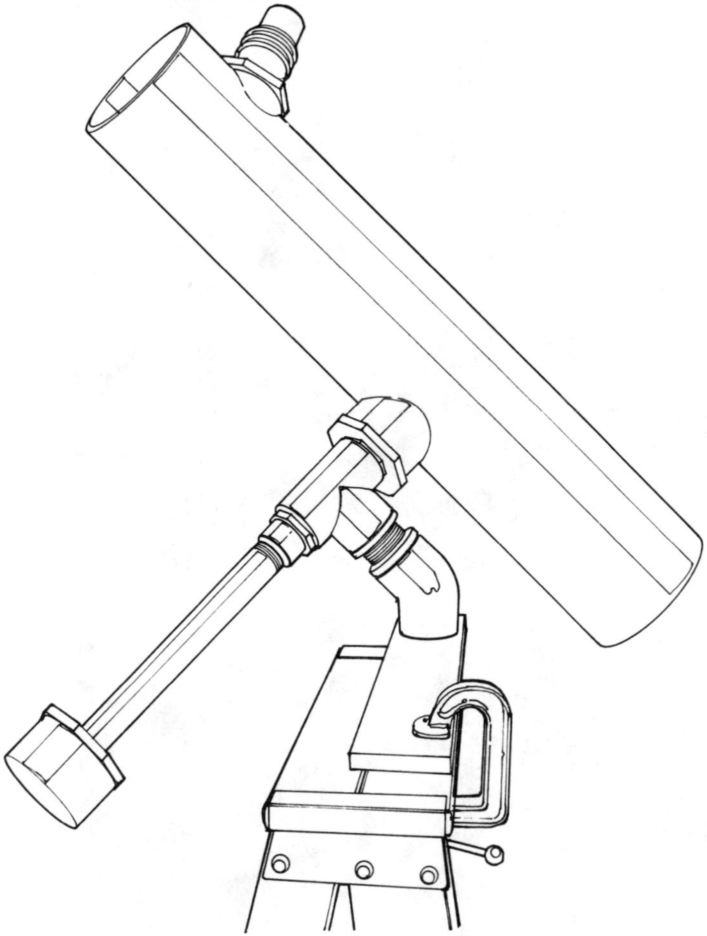

Reflecting telescope mounted on a stepladder.

Photograph of the moon taken through a three-inch telescope. *Courtesy Steven D. Freedman, Yale University*

Newton improved his own astronomical instruments, but he did not make that many observations of the stars himself. He was more concerned with using the accurate results of other people's observations as data to help him discover why the sun, moon and stars moved in the orbits explained by Kepler and Galileo.

Both Galileo and Kepler suspected that some force existed that explained their laws of falling objects and planetary motion. Legend has it that Newton discovered how the universe is held together—by means of gravity—while he was having tea in his garden in 1666 and an apple fell on his head! He realized that once the apple ripped away from the twig it was a free body in space just like the moon. Did the same force which kept the moon in its orbit also pull the apple to the Earth? He reasoned that if this force didn't pull the moon toward the Earth, then it would fly off into space. Newton's theory was that the force of gravity depends upon the amount of matter in the two bodies (such as the Earth and the moon) which are attracted to one another. Secondly, the force depended upon the distance between the two bodies.

For example, the force by which Earth pulls an elephant will be greater than that exerted on a mouse. This is because the elephant contains more matter. The pull of the Earth upon the elephant is called the weight of the elephant. Therefore, gravity is the reason the elephant weighs more than the mouse.

Newton attempted to prove his theory through a **formula** he developed. Since his calculations included an inaccurate measurement of the Earth's radius, his results were not completely satisfactory. He would not publish them for over 20 years! Throughout the years Newton spent at Trinity College as a professor, he held back many of his important discoveries until he was convinced they were completely developed and proven. In fact, he had already conceived his most important theories by the time he was 24 years old.

FORMULA

A mathematical statement, often an equation or a rule, principle, answer or other logical relation. In this case, Newton's formula was that every planet is attracted by the sun in direct proportion to its mass and in inverse proportion to the square of its distance from the sun. The same formula applied to bodies on Earth as well.

Years later when Newton obtained the correct figure for the radius of the Earth, he was able to complete his calculations and prove that the force of gravity did indeed extend to the moon. He was then able to conclude that all the heavenly bodies in the universe possessed strong forces that allowed them to attract one another. His notion of the universe also placed the sun, which was the largest body, in the center where it pulled all the others toward it in elliptical orbits.

As seemed to be his way of doing things throughout his life, Newton put his notes away and went back to other studies. Not until five years later, when the noted astronomer Edmund Halley came to Cambridge to visit him, was Newton pursuaded to try to find his calculations. Newton could not find them, so he agreed to recalculate them. Both Halley and another scientist, Robert Hooke, had used Kepler's law of planetary motions to conclude that the force of attraction between heavenly bodies must vary according to their size and the distance between them. But they could not prove their ideas mathematically. So Halley came to visit Newton to see if he had the answer. Halley was convinced that Newton's proof of the law of gravitational pull combined with Kepler's law of planetary motion would extend astronomical knowledge of the universe way beyond the planets that could then be observed.

Halley was also responsible for convincing Newton to publish his discovery and for paying for its publication in 1687. Entitled *Philosophiae Naturalis Principia Mathematica* (Mathematical Principles of Natural Philosophy), the work is usually simply called the *Principia.* The third part of the work which is called the "System of the World" described the unified system of the universe. A single law of force explained the motions of the planets, Earth, and even comets, as well as falling bodies and the rise and fall of the tides on Earth. Newton was the first astronomer and scientist to propose a theory of cosmology that could be proven mathematically. It was a long step from the purely philosophical theories of Aristotle. In Newton's system, motion was the natural state of every body, even on Earth. His laws of motion, including the law that to every action there is always an opposite and equal reaction, provide the foundations for the science of mechanics, a division of physics. These discoveries make it possible for modern people not only to design machines, but also to calculate how much energy will be required to complete a specific task.

Newton's *Principia* also helped reinforce Galileo's earlier belief in the value of rational experiments. It helped establish the experiment as the most effective way to study nature. He left Cambridge in 1696 to later become the Master of the **Mint** and was knighted in 1705. During the last years of his life, Newton was widely honored and very famous. Even though he made no great scientific discoveries during these years, he continued to quietly speculate on problems in science. These **speculations** helped inspire the lively discussions about science that were popping up with groups such as the **Royal Society of London**. Scientists were beginning to put their faith in reason and science, rather than the combination of science and religion practiced during the Middle Ages. It was also realized, more and more, that science need not disagree with religion. Newton himself was a very religious man. He believed that his system of the universe worked like a beautiful machine and that a rational God had made it and had started the machine. Newton showed that the best way to understand such a system was to use reason in asking questions—questions that could be answered through observation and experiment.

Since mirrors were very difficult to grind in the 1600s, most astronomers continued to use a refracting telescope instead of Newton's reflecting telescope. Many of these instruments were based upon Kepler's descriptions of a telescope. They had a convex eyepiece placed behind the focal plane to greatly increase their field of view.

MINT

The place where the coins of a country are manufactured by the government.

SPECULATIONS

Careful thought, a reflection. A guessing.

ROYAL SOCIETY OF LONDON

The Royal Society of London for improving natural knowledge was set up in 1662. Its purpose was to encourage professional and amateur scientists to pursue scientific studies.

An Arabic astronomer in the 1600s with a six-foot-long refracting telescope.

The image seen through the eyepiece was inverted because it was formed from the objective lens at the focal plane. When another lens was added, the image was reinverted, or right-side up. The Keplarian telescope could magnify up to 1,000 times. It was also possible to place two thin pieces of wire at the focal plane to help bring the image of a star into focus at their point of intersection. This device aided astronomers in determining the positions of the stars with greater accuracy than with an open sight.

The main problems faced by astronomers at the time were chromatic and spherical distortions. Spherical distortions appeared because rays that passed through the middle of the lens were brought to focus at a different point than those that passed through the edge of the lens. They tried building telescopes with very long focal lengths to reduce these distortions. Some of these instruments were up to 200 feet long. They had to be supported by ropes and pulleys. Johan Hevelius, a German astronomer, built one of these huge telescopes that had a focal length of 150 feet!

Johan Hevelius (1611-1687)

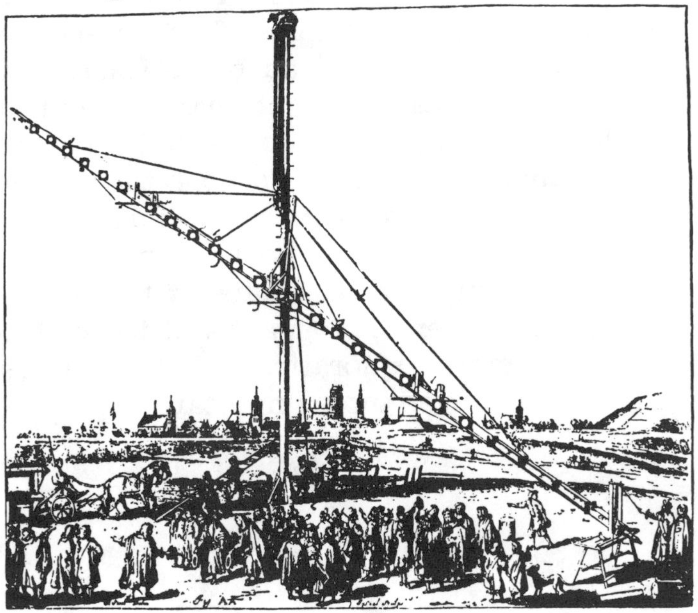

Hevelius' 150-foot telescope

SOLAR SYSTEM

The sun and nine planets, together with the moons, comets and other celestial bodies that revolve around it.

GALAXY

Group of billions of stars, gas and dust which form one system in the universe. The Earth and sun are part of one galaxy, the Milky Way. The word "galaxy" comes from the Greek word *galaktos*, which means "milk." The Greeks must have thought that a cluster of stars looked like a splash of milk across the sky.

Hevelius also published detailed accounts of the instruments used at his time. He gave directions for building both Galilean and Keplerian telescopes. His most important contributions, though, were probably his discoveries about the stars and planets in the **solar system** and **galaxy**. Hevelius published two large books on comets. They recorded the appearances of all known comets. He was also responsible for mapping the moon. Mountains, valleys, craters and other features that he observed were named after those on Earth. Thus, the moon has alps and even a Pacific Ocean!

By 1651, an Italian Jesuit, Joannes Riccioli, published another map of the moon. On his map some features were given romantic names like the "Sea of Tranquility." Others were named after philosophers, scientists and astronomers. Riccioli's map has been revised again and again as better telescopes have come into use. Today, the moon is still being mapped and named. Photographs of its surface are being taken from satellites traveling in space and by astronauts who have explored its surface. Apollo 12 astronauts named a cluster of craters, "Snowman." Other Apollo mission astronauts have named a dangerous area "Spook," and another after Cochise, the Apache. Finally, it was decided that all features on the moon should be named after great scientists, American astronauts and Soviet cosmonauts, or with miscellaneous male and female names. Maps of the moon will become even more detailed as better photographs of its surface are taken. And someday, when more explorations of the universe are made, maps of the moon will be even more important.

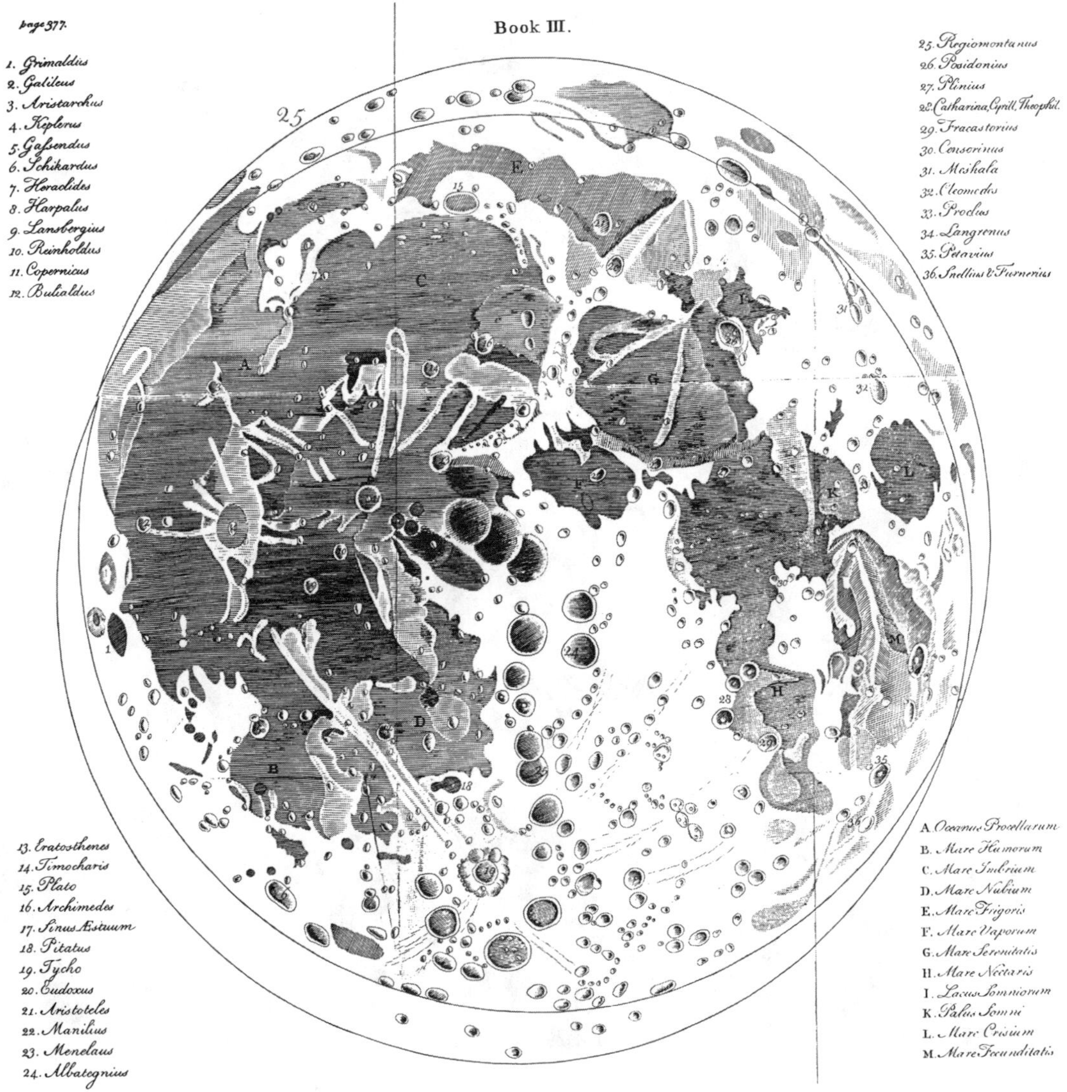

Riccioli's map of the moon which was republished in Roger Long's *Astronomy in Five Books* in 1785. Note that 11 is the crater Copernicus.

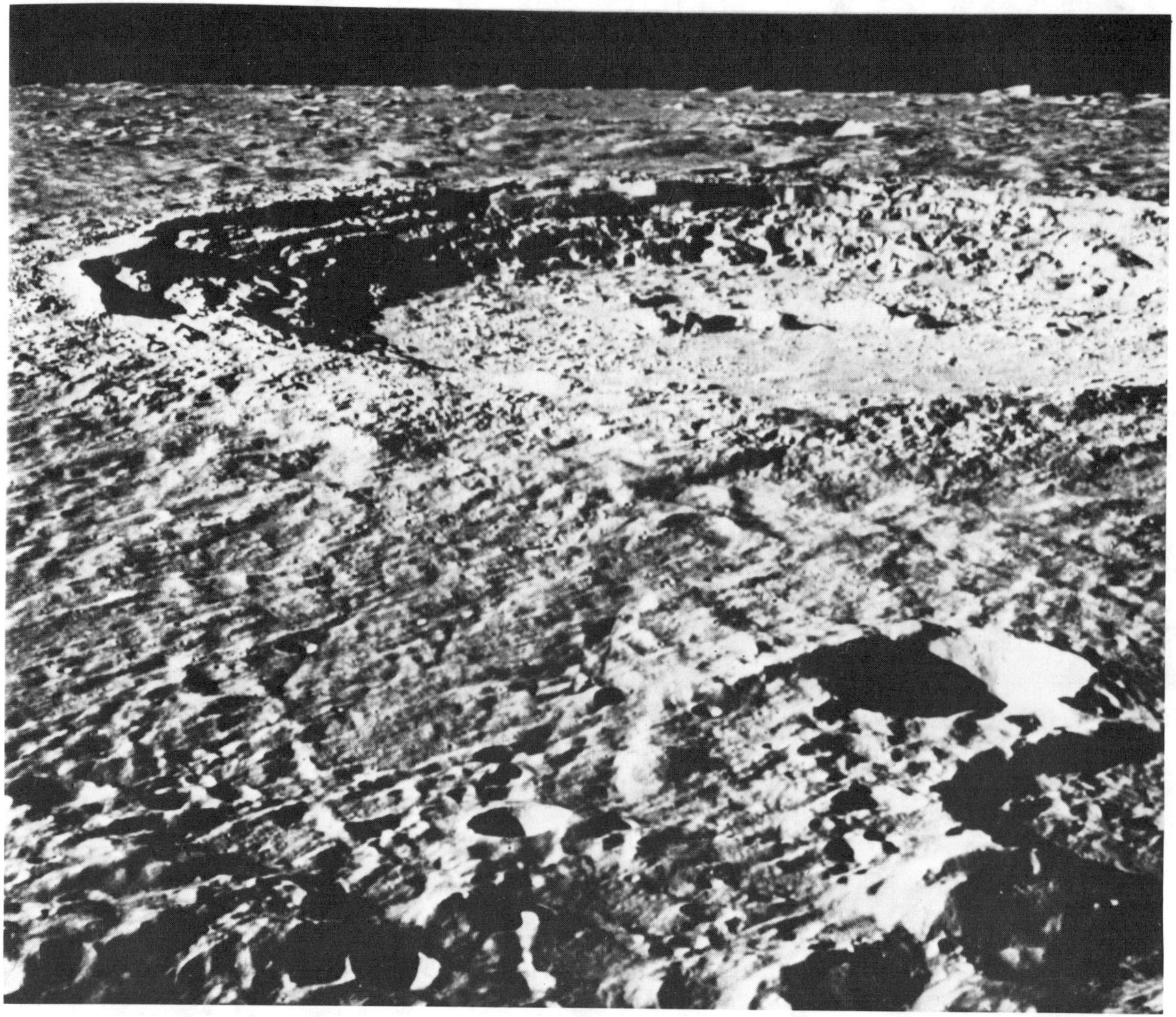

The large crater on the moon's surface known as Copernicus. This photograph shows how the crater appeared from the Apollo 17 spacecraft, looking southward over the Carpathian Mountains. *Courtesy of the National Aeronautics and Space Administration*

Edmund Halley was another of the astronomers during Newton's time who made important contributions to our knowledge of the universe. During the late 1600s he went to St. Helena, an island in the Atlantic Ocean. Using a telescope, Halley made careful observations of the sky over the Southern Hemisphere from the island. They were published as a star catalog. When he returned to England he began to compile a complete record of every sighting of comets known to mankind. By using Newton's law of gravitation, Halley was able to prove that comets are part of our solar system. They move in elliptical orbits and are visible when they pass close to the sun.

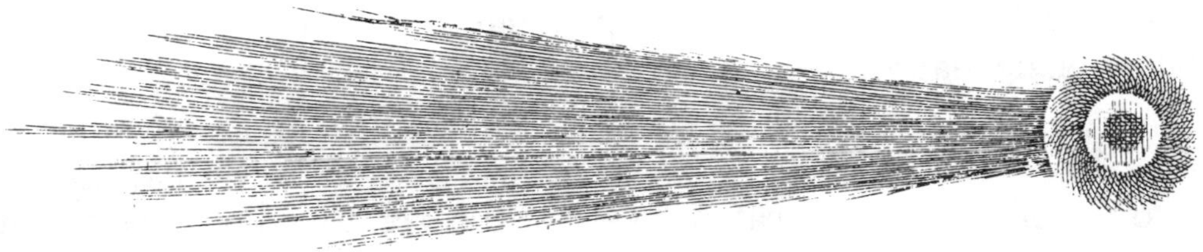

The comet of 1661, which appeared as an illustration in Roger Long's *Astronomy in Five Books,* published in 1785.

Halley's records of periodic comet appearances enabled him to notice similarities in some of their orbits. He suspected that these similarities represented the same comet. Thus in 1682, when one of these comets appeared again, Halley was able to predict that its next appearance would be 77 years later in 1759. He did not live to see that his prediction was correct for the comet that still bears his name. Halley's Comet was first seen in 240 B.C. and since it reappears every 75-77 years, should be seen again in 1986.

Drawing based upon the illustration of Halley's Comet in the Bayeux Tapestry, which was made in England during the 1100s.

MERIDIAN

Imaginary circle around the Earth passing through the North and South Poles. Most of the world bases its time on "GMT" or Greenwich Mean (solar) Time, which is determined by setting noon as the time when the sun is directly overhead the Greenwich meridian.

Even in Halley's time, many people were still very much afraid of comets. Ancient people thought that their appearance indicated that the gods were angry. Throughout early history all sorts of disasters seemed to follow the appearance of a comet.

Later in his life, Halley was appointed the Astronomer Royal at the Royal Observatory in Greenwich, England. He collected new telescopes and other instruments for the famous observatory which had been founded in 1675. It was designed by the architect, Sir Christopher Wren, so that the zero **meridian** runs through the center of the main large octagonal room. One of the main purposes of the observatory is to assist in navigation by providing information on tides. The Astronomer Royal observes the motions of the sun, moon and other planets as well as compiling exact star charts. Astronomers such as John Flamstead (the first Royal Astronomer), Newton and Halley made many important observations there.

Throughout the late 1600s many people were interested in building large observatories. One of these was built in Peking, China. Located on top of one of the old watchtowers of the great wall of the city, the Chinese Imperial Astronomical Observatory included: a zodiacal sphere, an equinoctial sphere, an azimuthal horizon, a quadrant, a sextant, and a celestial globe.

One of the people with fantastic plans for constructing huge astronomical instruments and buildings to house them was Giovanni Cassini, a professor of astronomy at the University of Bologna. In the 1660s he accepted an invitation from the French king, Louis XIV, to come to Paris. He was to be the director of an observatory which was being built for the Academy of Sciences by the architect Perrault.

An illustration of the 12.8-inch equatorial telescope at the Greenwich Observatory in about 1860. From E. Dunkin, *The Midnight Sky*, (1880).

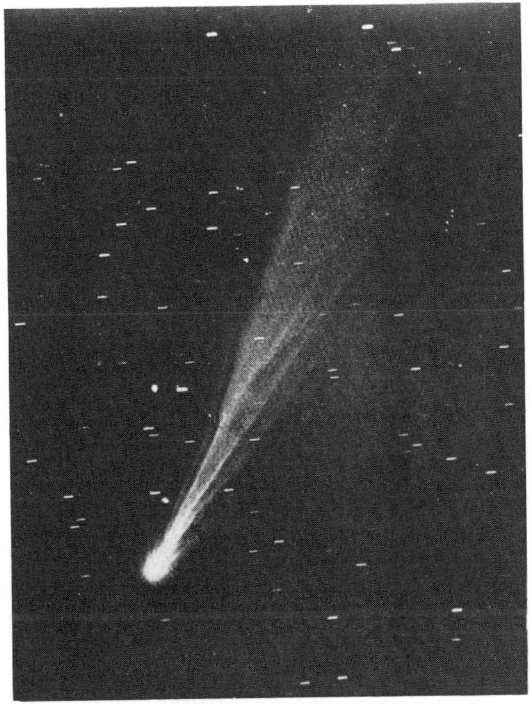

Photograph of the Morehouse Comet, taken at the Greenwich Observatory in 1908.

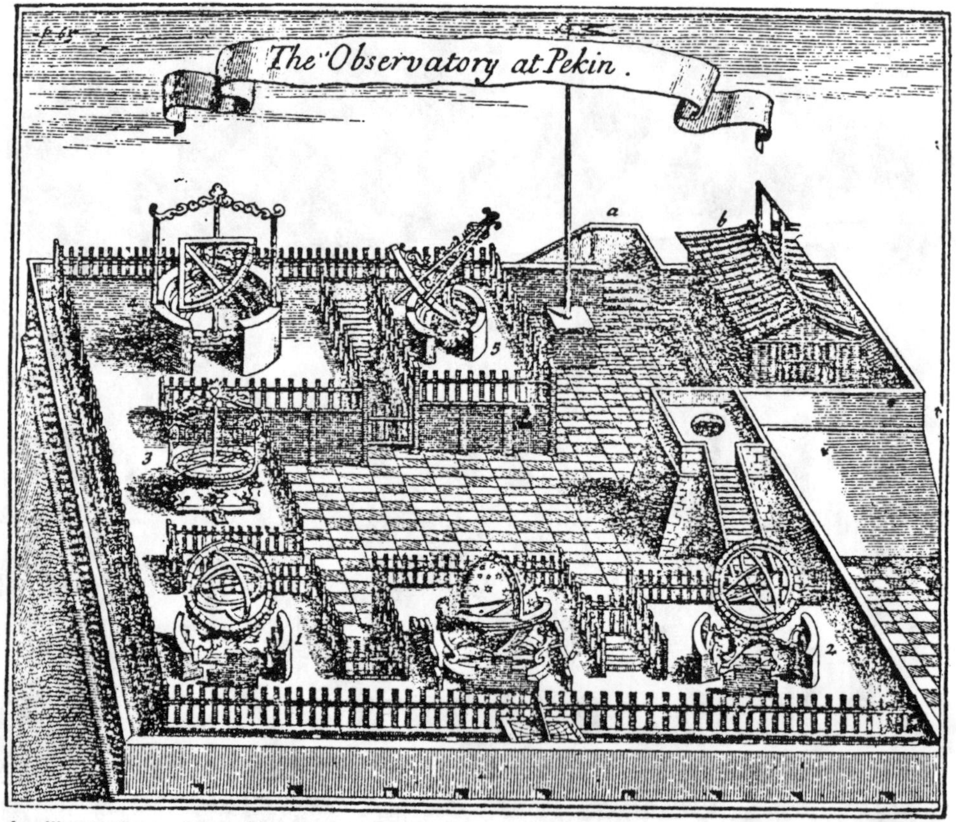

An illustration of the Chinese Imperial Astronomical Observatory from Le Comte's *Voyage to China,* (1698).

Although the building was completed in 1671, it did not work very well as an observatory. So Cassini set up some of the instruments he had brought with him outside. He discovered the division of the rings of the planet Saturn, which still bear his name today. The observatory itself was used by many scientists and astronomers, not just to make observations, but as a meeting place to compare their notes and discuss the results of their experiments.

An illustration of the great equatorial telescope set up at the Paris Observatory during the late 1800s.

Some of the greatest advances in observational astronomy were made during the late-1700s and early 1800s by Sir William Herschel. Trained as a musician, Herschel began his study of astronomy as a hobby. He found most of the telescopes available too small or too expensive, so like many other famous astronomers, he built his own.

Great advances were being made in the construction of the telescope during this time. In 1752, a London optician, John Dolland, had almost totally eliminated spherical and chromatic aberrations in refracting telescopes. He accomplished this by making an objective

Sir William Herschel (1738-1822)

lens from a combination of two lenses. One was a double convex lens made from crown glass. The other was a double concave lens of flint glass. The different refractive qualities of these two types of lenses cancelled out each other to create an achromatic lens.

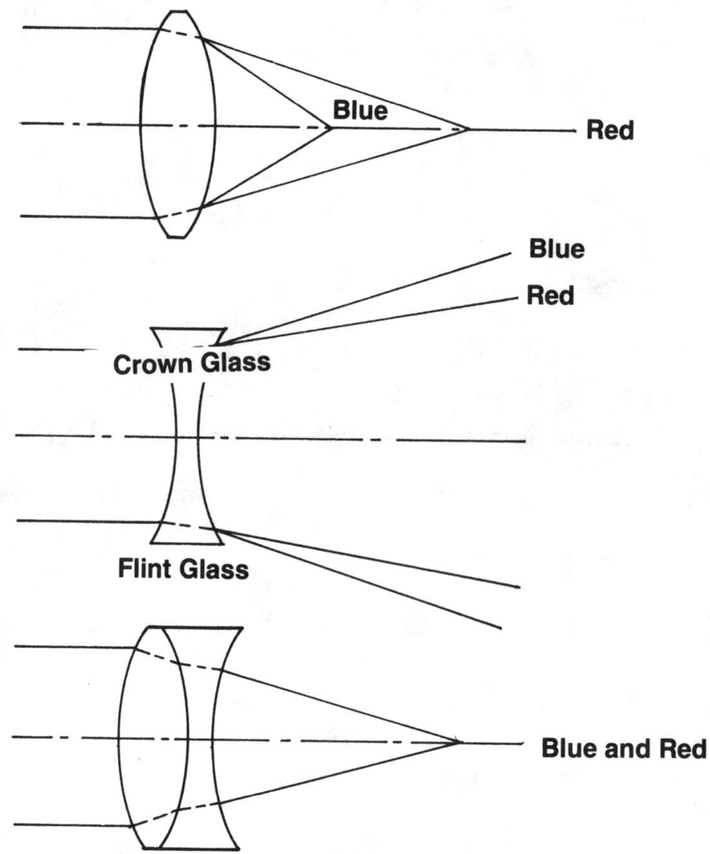

Combination achromatic lens

When Herschel took up astronomy in 1775, though, achromatic telescopes were still too expensive and too small for him to use. Instead, he built a Newtonian reflecting telescope. By the mid-1700s, it was possible to easily grind concave mirrors for this type of reflecting telescope. It became the favorite instrument of most astronomers.

Throughout his own career, Herschel usually used 7 to 20-foot telescopes with mirrors of between 6 and 19-inches diameter. However, he did build several huge telescopes. One, which was completed in 1789, had a length of 40 feet with a 48-inch mirror. The mirror was made of speculum, a copper-tin alloy which many astronomers used before they had learned to cast large mirrors of glass. Herschel had to raise and lower the huge instrument with pulleys. Levers made it possible to move it from side to side. A circular rail allowed it to be turned in order to observe different sections of the sky. Even though the larger mirror allowed more light to be collected by the telescope, there were too many problems with its size and maintenance to make it a very practical instrument for Herschel to handle.

Herschel made his observations with the goal of expanding our knowledge of the construction of the heavens. His observations allowed him to enlarge our known solar system by discovering the planet Uranus in 1781. The planet had actually been observed earlier at Greenwich by such astronomers as Flamstead, but they all had thought that it was an ordinary star. Herschel himself had at first thought that it was a comet.

Herschel's sister, Caroline, was a partner in his astronomical studies. She lived with him until his marriage, when she moved close-by. Caroline helped grind and polish mirrors, record and calculate observations. She became one of the first women astronomers, discovering more than six comets on her own.

Caroline Herschel at age 92

Herschel's giant 40-foot telescope

Herschel and his sister discovered and cataloged several hundred double stars that held together through mutual attraction. They also cataloged over 2,500 **nebulae**. Herschel believed that many of these were actually collections of stars, or what we call star clusters today. Through their systematic observations, the Herschels were also able to discover that our galaxy, the Milky Way, is shaped like a flat disc with the sun close to its center. Today, we know that the galaxy is actually shaped more like a fried egg with a bulge in the center. As they probed deeper into space with their telescopes, the universe became more complex. Herschel concluded that our solar system was one of motion, whose disturbances did not destroy the system, but were contained within it.

NEBULAE

Vast dark or bright cloudlike clusters of stars or masses, or dust particles and gases. Today astronomers use the term "galactic nebula" for the clouds of dust or gas that derive their radiation from the stars within them.

One of the scientists to discover that the disturbances astronomers had observed in the solar system were compensated for within the system itself was the Frenchman Pierre Laplace. Since Newton's law of gravity applied only to two bodies, Laplace and other scientists at that time were concerned with how more than two bodies interacted. Laplace's great study on celestial mechanics provided a mathematical solution to the problem of the third body. He also described a stable solar system that had existed for an incredible period of time. Laplace believed that the system began when rings broke off from a hot, rotating mass of gas and formed the planets.

Pierre Laplace (1749-1827) *Culver Pictures*

Throughout the 1700s, astronomy was flourishing both in Europe and in the American Colonies. Many of the important contributions being made to the science were being made by mathematicians who were also astronomers. One of these people was John Winthrop, a Professor of Mathematics, Natural and Experimental Philosophy at Harvard College. One of Winthrop's telescopes, a **Gregorian** reflector, was made by James Short, an instrument maker in London in 1740.

GREGORIAN

Reflecting telescope invented by James Gregory. The concave primary mirror reflected light to a small concave secondary mirror. The light then traveled down the telescope tube and entered the eyepiece through a hole in the center of the primary mirror.

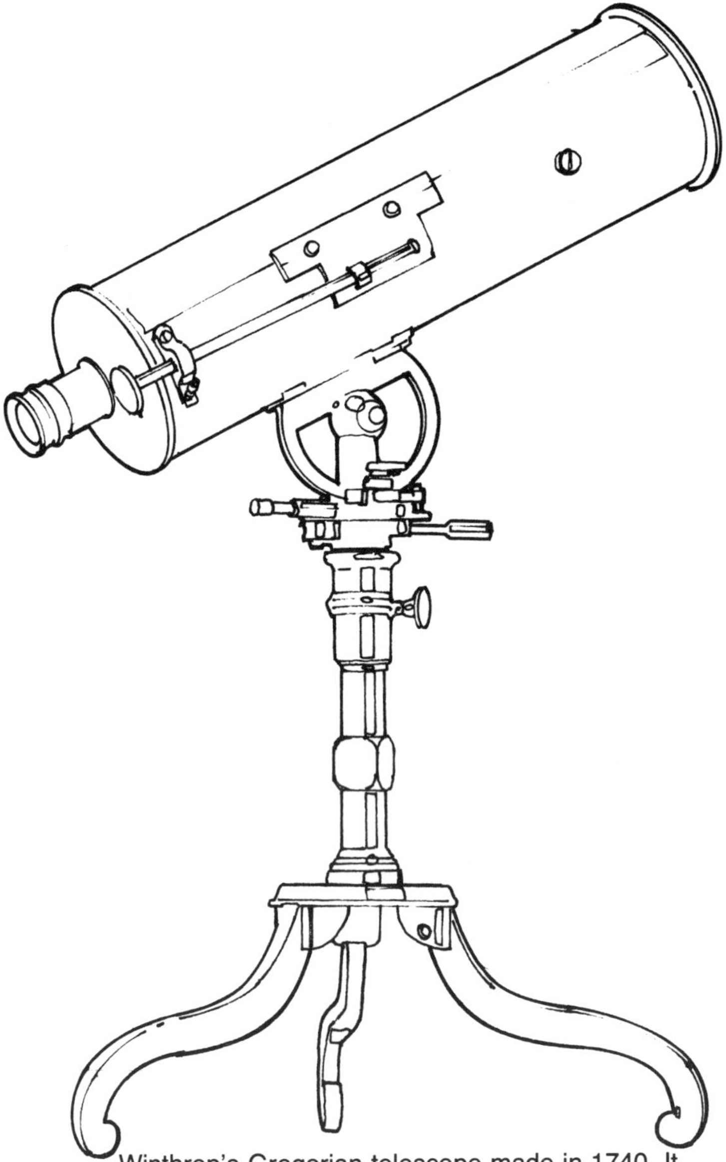

Winthrop's Gregorian telescope made in 1740. It is all brass with a barrel of 12″, diameter of 3¼″ and height of 17½″.

Another important early American astronomer and mathematician was Benjamin Banneker (1731-1806). Banneker was a self-taught, black tobacco planter in Baltimore County, Maryland. The freed son of a slave, his interest in astronomy was encouraged by George Ellicott, a mill owner who lived nearby him. Ellicott gave him his first pedestal telescope and books on astronomy. Although he began his astronomical and mathematical studies late in life, Banneker eventually served as a scientific assistant in the survey of the District of Columbia, where the new nation's capital was being planned. One of the instruments he used for the survey was an achromatic telescope with a pedestal, which was made by John Dolland of London. Banneker was also an almanac maker. He was 60 years old when his first almanac was published in 1792. Entitled, *Benjamin Banneker's Pennsylvania, Deleware, Maryland and Virginia Almanac*, it was compiled by means of his own calculations and observations. Included in the almanac were: diagrams of projected eclipses, times of sunrise and sunset, positions of the moon in relation to the constellations each month, and weather and seasonal predictions.

Other astronomers were trying to open up the universe even further by building even larger telescopes. The Earl of Rosse, an Irish nobleman, built an even larger telescope than Herschel's at Birr Castle in Ireland. For many years it was the largest telescope in the world. The mirror weighed four tons. It had a six-foot diameter and a focal length of 54 feet. Because of its huge light gathering capacity, Lord Rosse was able to study Herschel's nebulae. Although he was the first person to see that some of them were spiral in form, he still could not distinguish for sure whether they were made up of stars or just cloudy objects.

An engraving from the late 1800s of Lord Rosse's telescope

By the end of the 1800s, most of the very large telescopes and observatories were being built by official institutions. Photography had also become an extremely important astronomical tool. As early as the 1840s, photographs had been taken of the moon at the Harvard College Observatory. In fact, today most astronomical studies depend upon photography and most of the world's greatest telescopes are equipped to take photographs.

Three of the world's greatest observatories—Yerkes, Mt. Wilson, and Palomar—were built as a result of the dreams and efforts of one American, George Ellery Hale. Even as a young boy, Hale was fascinated with astronomy. Hale himself wrote:

> As a boy, largely through the constant encouragement of my father, I became interested in tools and machinery at a very early age; I always had a small shop with tools, first in the house, and later in a building of my own construction in the yard. I also had a little laboratory where I performed simple chemical experiments, made batteries and induction coils, worked with a microscope, etc. After construction of a small telescope for myself, my father bought me an excellent 4-inch Clark. I used this constantly, but my enthusiasm reached the highest pitch when I learned something about the spectroscope. My greatest ambition was to photograph a spectrum and this I soon succeeded in doing with a small one prism spectroscope purchased for me by my father. I think this was in 1884. Solar spectroscopic work appealed to me above all things and I read everything I could find on the subject. My father always bought for me any books that I needed, but in the case of instruments his policy always was to induce me to construct my own first and then to give me a good instrument if my early experiments were successful. In 1888 he built for me after my own designs, a spectroscopic laboratory in which a Rowland concave grating of 10 feet focal length was erected. This was the nucleus of the Kenwood Observatory.

George Ellery Hale (1868-1938)

An illustration of a homemade spectroscope which was used as a model by Hale. From Cassell's *Book of Sports and Pastimes,* 1881.

SPECTROSCOPE

An instrument for producing and examining the spectrum of a ray of light.

Hale's tools were the telescope, the **spectroscope** and the photographic plate. At first he worked in small observatories such as Kenwood. By the age of 23, Hale was famous for inventing the spectroheliograph, a special instrument used to take spectroheliograms, or photographs of the sun. Spectroheliograms are taken on a narrow wavelength band that is centered on a select wavelength. In

Some of the instruments at the Kenwood Observatory

The Kenwood Observatory behind the Hale home in Chicago where a 12″ refractor telescope and spectroscope were set up in 1891.

1892, Hale was appointed as Associate Professor of **Astrophysics** at the University of Chicago. He soon wanted to build a much larger telescope, with at least a 40″ refractor, to study the sun. Hale finally persuaded a very wealthy Chicago businessman, Charles T. Yerkes, to put up over $30,000 to finance the Yerkes Observatory. By 1897 it was completed at Williams Bay, Wisconsin, about 70 miles from Chicago. The objective lens of the telescope was 40″ in diameter. The tube was 62″ long. Yerkes was not just an observatory, but a huge laboratory where astronomical, spectroscopic, photographic and optical studies were pursued.

ASTROPHYSICS

The branch of astronomy concerned with the physical and chemical makeup of the heavenly bodies.

Yerkes Observatory from the northeast

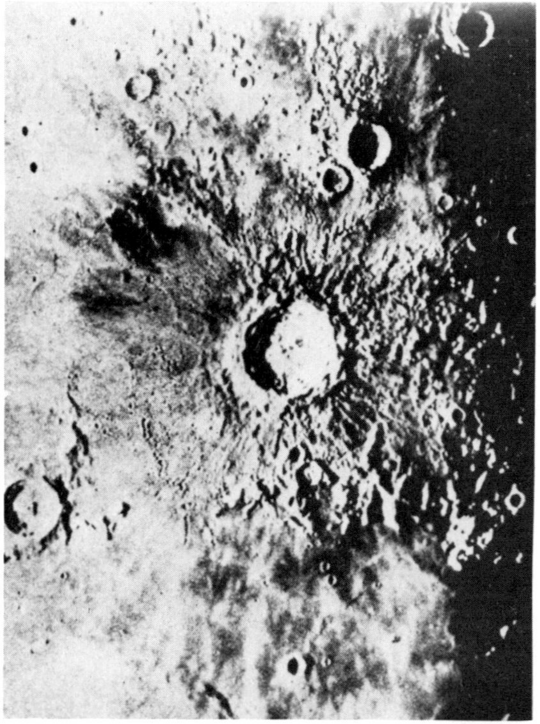

Photograph of the moon crater Copernicus, taken with the 40″ refractor at the Yerkes Observatory.

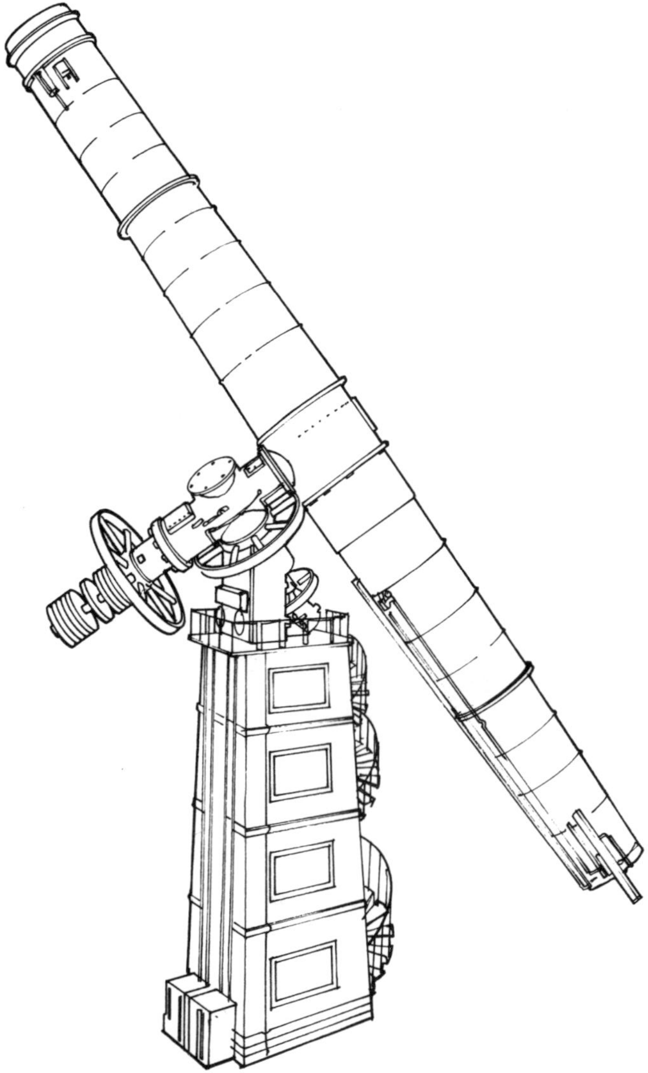

The 40″ refractor telescope at the Yerkes Observatory

Hale was not completely satisfied with the amount of light provided by the Yerkes refractor. He wanted to explore even more remote solar systems. Again he found a wealthy businessman, Andrew Carnegie, who was willing to finance a 60″ reflector telescope. It was built at Mt. Wilson in California in 1908. By 1917, a second, even larger reflector

One-hundred-inch reflector at Mt. Wilson Observatory

Photograph of the spiral nebula in the Great Bear Constellation, taken with the 60″ reflector telescope at Mt. Wilson, California.

ANDROMEDA

Constellation in the Northern Celestial Hemisphere. It is about 2.2 million light years away. Andromeda has about 100,000 million suns.

of 100″ was also built on Mt. Wilson. With this huge reflector, astronomers were able to take photographs of the great nebulae in **Andromeda**. The pictures proved Herschel's and Laplace's theories that these nebulae were made up of stars. In fact, these nebulae are really faraway galaxies. The large Mt. Wilson reflector is capable of viewing over 100 million galaxies! Probably Hale's greatest contributions to astronomy were his dreams of more powerful instruments of observation and his ability to successfully get them built.

Even though he had built three of the world's largest telescopes, he still dreamed of building an even larger one. Also a reflector, but this time with a reflecting mirror 200″ in diameter. This telescope was completed on Mt. Palomar in 1948. Unfortunately, Hale died in 1938 before the largest reflector on Earth was finished. Today it is named the Hale Reflector. In 1970, the Mt. Wilson and Palomar Observatories were renamed Hale Observatories. The Hale reflecting telescope is so large that the astronomer must sit in a cage slung from its side. The instrument's chief function, though, is to take photographs which are then studied by astronomers.

The Hale Observatories on Mt. Wilson and Mt. Palomar became famous centers for the study of astronomy and physics. Observations made there provided the data for the development of totally new theories of the universe. In 1913, for example, the physicist Albert Einstein wrote to George Hale from Zurich, Switzerland. Einstein wanted some information on observations made with the Mt. Wilson telescope that would help prove his theories of relativity.

Einstein was a theoretical physicist who only needed books, paper and pen to develop his theories. Often his ideas were based on the data that had been obtained experimentally by others. Throughout his life he was seeking one uniform law that governed all things in the universe.

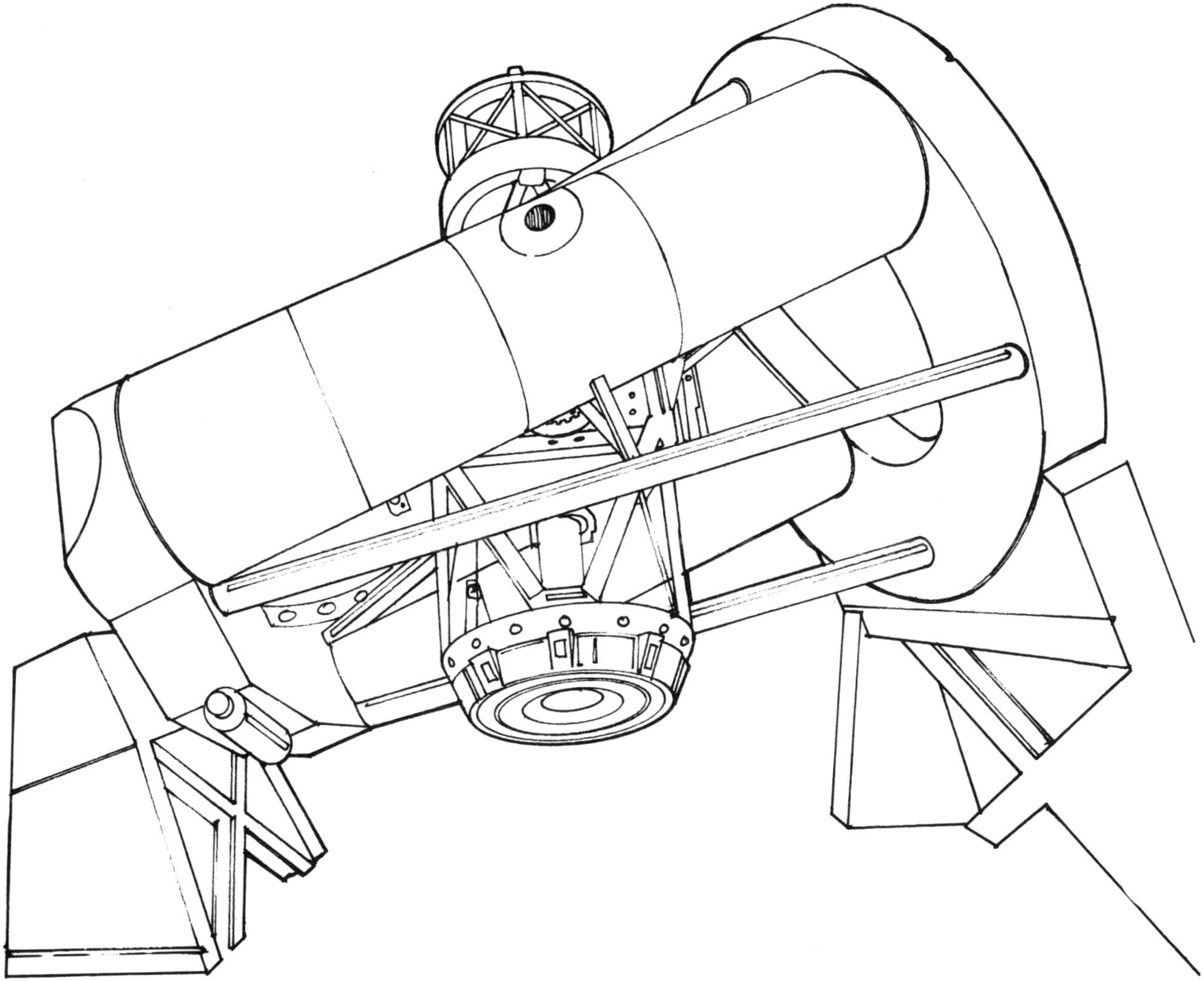

The huge reflecting telescope at the Palomar Observatory. The 200-inch mirror is in the round basket slung from the center of the telescope.

Clusters of galaxies seen at the outer range of the 200-inch reflecting telescope on Mt. Palomar. The galaxies are seen as they existed one billion years ago. Since they are so far away, it would have taken their light that long to travel to Earth, even if it traveled at the speed of 186,000 miles per second, the maximum speed of light. *Courtesy of Mount Wilson and Palomar Observatories*

Even though Newtonian physics worked fairly well on Earth, it did not help astronomers understand the enormous masses in space that sometimes travel close to the speed of light. A more comprehensive theory of the relationship of bodies and energy was needed. Two main questions were bothering physicists and astronomers at that time. Since light waves seemed to be able to travel through a **vacuum** without any supporting material, they wondered if a vacuum, or an **ether** existed or not. Secondly, they wanted to know whether or not the speed, or velocity of light, is dependent upon the speed of the observer.

VACUUM

An empty space without even air in it.

ETHER

The upper regions of space that are beyond the Earth's atmosphere. The clear sky, the heavens.

In 1905 Einstein developed his Special Theory of **Relativity**. Its two main **axioms** were: 1) that ether cannot be detected and that the velocity of light is the same for all observers; and 2) that all motion is relative, so no observer is absolutely at rest. The first axiom applied to astronomy in the following way. When astronomers examine double stars through a telescope, the speed of light is the same from each star. This is true even though one star is moving toward the Earth and the other may be moving away from it. This fact is one of the proofs that the speed of light is independent of the speed of the source of light and is always 186,000 miles per second.

Newton had considered space and time to be absolute. In his system of physics there would be no differences between different points in space. For example, parallel lines would never meet. Einstein's second axiom, though, states that no observer is absolutely at rest. An astronaut in a spacecraft traveling away from the Earth along the Earth's orbit would see the Earth as moving away. He would assume that he was stationary. To an observer on Earth, who assumed himself to be stationary, the spacecraft would appear to be moving away. To an observer in space, both the Earth and the spacecraft would appear to be moving. Einstein's theory, therefore, states that velocities can only be calculated in terms relative to the observer.

Perhaps it is easiest to understand this axiom if you pretend that you are standing on a road watching two trains speed by. One train is traveling at 60 miles per hour. The other is only going by at 40 miles per hour. As a bystander you can determine the differences in the speeds relative to your position, but to the engineer of the slower train, the faster train is traveling past him at only 20 miles per hour. Therefore, we say that the faster train has a relative velocity of 20 miles per hour.

RELATIVITY

In this case the word "relativity" refers to the relative velocities or speeds.

AXIOM

Statement that is taken to be true without proof.

The Special Theory of Relativity makes it possible for us to better understand what happens to length, mass and time in an object or system that is moving relative to the observer. For example, if you had been able to see the 60 mile per hour train stopped at a station, it would have seemed to be longer than when you observed it traveling at 60 miles per hour. In the train itself, mass would have increased as the speed increased. Time would also pass more slowly than for the observer on the road. This theory also applies to space travel. For this reason it is possible for a spacecraft to leave Earth, and after traveling at very fast speeds, to return to Earth with less time elapsed on the spacecraft than has elapsed on Earth.

Another important part of Einstein's Special Theory of Relativity is that you can make matter disappear and energy appear in its place. This is because mass and energy are equivalent, or they have the same value. This relationship is expressed in the famous equation, "$E=mc^2$." E is the energy; m is the mass; and c is the velocity of light. The equation makes it possible to calculate the total amount of energy in any given mass. It can also be used by astrophysicists to try to discover the sources of energy for the stars. In fact, they have learned that the energy of the stars is the result of **nuclear reactions** in their interiors.

NUCLEAR REACTION

A reaction that changes the energy, makeup or structure of the atomic nucleus or the central part of the atom.

Einstein's General Theory of Relativity, which was introduced in 1916, also provided physicists and astronomers with new ideas on space and time. It is basically a theory about gravity. Isaac Newton had described gravity as a force. Einstein said that it is more like a curvature in space. Therefore, the Earth maintains its orbit, not because the force of the sun holds it in place, but because it is always following the easiest pathway through the solar system created by the sun. According to the General Theory of Relativity, the light of stars would be deflected as it passed through the sharply curved portion of space near the sun.

Albert Einstein (1879-1955) riding a bicycle during his visit to the Mount Wilson Observatory in the 1930s. *Courtesy of the Archives, California Institute of Technology*

In 1913, Einstein had written to Hale about using the huge Mt. Wilson telescope to provide information on this aspect of his theory. He wanted to know whether the gravitational force of the sun would deflect the light of a star that passed by it. He also needed to know how close to the sun bright stars could be seen in daylight. Even though the astronomers at Mt. Wilson were unable to provide all of the answers Einstein wanted, he was able to complete the theory three years later. The theories of relativity have helped scientists work out new models of the overall structure of the universe. Since the theories were presented in the early 1900s, astronomical observations and experiments in laboratories have confirmed many of their basic axioms.

Einstein visited Mt. Wilson during the 1930s. Both he and his wife were shown the giant reflecting telescope. There is a wonderful story about their visit which points out the differences between the theoretical and experimental approaches to discovering the secrets of the universe. It is said that Mrs. Einstein asked why the telescope was so big. The director of the observatory told her that they were trying to establish the shape and makeup of the universe. She replied that her husband worked on the same problem, on the back of an old envelope!

New advances in astronomical instruments, such as the Hale Observatory telescopes, have made it possible to experimentally test many different theories about our universe. When Galileo looked through his first telescope in 1610, for example, a new universe was revealed to him. The telescope expanded his ability to perceive and understand the heavens. Saturn and the four moons of Jupiter were seen for the first time. A revolution in man's understanding of the universe had begun. Old theories were questioned as a result of the discovery and use of the telescope.

This revolution is continuing in our own time through the design and use of new types of telescopes and instruments to help us understand the planets, stars and the universe. Among the most exciting of these new instruments is the **radio telescope**. The radio telescope does not form an image of the planet or other object at which it is pointed, as an optical telescope does. Instead, it collects the radio waves from that object on a detector so that their intensity can be recorded. The fact that a radio signal is picked up proves to us that something is out there in what would otherwise be the invisible reaches of outer space. Radio telescopes also serve as thermometers because the strength of the waves they pick up can be used to calculate the temperature of an object.

RADIO TELESCOPE

A sensitive, radio-antenna system which is used to gather and concentrate radio waves from objects in outer space, just as an optical telescope gathers and concentrates light waves.

The first radio telescope dates back to the early 1930s. In 1932 an American named Karl Jansky was asked by his employer, the Bell Telephone Laboratories, to find out what caused the serious static disturbances in transatlantic and ship-to-shore radio connections. Jansky built a huge rotating antenna to try to discover the source of the radio static. After collecting a great deal of data, Jansky concluded that some of the static that was interfering with the radio transmissions was probably coming from somewhere deep in our galaxy. It seemed to rise in the east and set in the west with the sun. Without realizing it, Jansky had made the first observations with a radio telescope and opened up the field of **radio astronomy**.

RADIO ASTRONOMY

The branch of astronomy in which the heavens are observed with a radio eye instead of an optical eye. It is concerned with the analysis of radio waves that are transmitted by heavenly bodies.

Grote Reber, another American who was an amateur radio broadcaster and radio engineer, soon learned of Jansky's discoveries. During the mid-1930s he constructed a thirty-one foot dish-shaped antenna in his backyard in Wheaton, Illinois. Reber began to record **radio waves** which his radio telescope collected.

RADIO WAVES

Form of energy which radiates from stars and other sources in outer space that cannot be seen or felt. They can only be detected with radio equipment.

Karl Jansky's huge "merry-go-around" antenna, built to find the source of static that was interfering with transatlantic radio communications. The platform is 100 feet long and 10 feet wide. It was mounted on four old Model *"T"* Ford wheels.

INTERSTELLAR

Between the stars.

Both Jansky and Reber were recording radio waves generated by clouds of **interstellar** gas and dust outside of the solar system. Using the antenna of their radio telescopes to capture these waves, they were able to convert them into electrical signals. By measuring the strength and other characteristics of the signals, astronomers began to realize that it would be possible to determine a great deal about the universe with radio telescopes. As these instruments swept the universe, it was possible to "eavesdrop on space."

Not only can the huge dish-shaped antenna of radio telescopes be used to collect signals from outer space, but some of them can also send waves into space. As **radars**, they then listen for the echoes of these waves reflected off anything they have hit. By timing these reflected waves, radio astronomers are able to learn how far away objects are and also something about their surface features. Radar astronomy can be used to study most of the closer heavenly bodies such as the sun, nearby stars, planets, comets and **meteors**. Transmissions sent to some of the stars in the galaxy take years for the signal to return because they are so far away.

Enormous advances have been made in the field of radio astronomy since World War II. Installations of telescopes with huge reflecting dishes have been built around the world. Since clouds or sunlight do not affect the wavelengths being received, these telescopes can operate in any weather, day or night. In Sorocco, New Mexico, a huge Y-shaped set-up of radio telescopes is being assembled to form the National Radio Observatory. By using computers it will be able to combine the data collected from each of the individual antennae, creating a very sensitive single radio telescope. The 27 giant antennae will be able to gather more radio waves and more clearly concentrate and analyze them than any other radio telescope. Combined, the antennas will operate as though they are one radio telescope, 17 miles in diameter. With this instrument, astronomers will be able to "see" to the very limits of the universe.

Radio astronomy is forcing astronomers to redefine their understanding of the universe. For example, we now know that every single particle of matter in the universe sends off energetic electrons at radio-wave frequencies. One of the strangest discoveries made with radio telescopes were **quasars**. A quasar is a starlike source of radio waves. Astronomers calculated that some of them may be 10

RADAR

Radar locates an object and measures the time it takes a radio wave to travel to a target and back. Thus, it measures time, not distance. For example, radar has been used to give us a precise measurement of the distance from the Earth to Venus.

METEORS

Lumps of metal and rock which enter our solar system at speeds up to 45 miles per second. When they enter the Earth's atmosphere their speed causes them to become very hot because of friction with the air. Usually they are vaporized. After they are vaporized, a trail of electrified air particles are left behind. It is this trail that has traditionally been called a "shooting star."

QUASARS

An abbreviation for quasisteller object which literally means "something like a star."

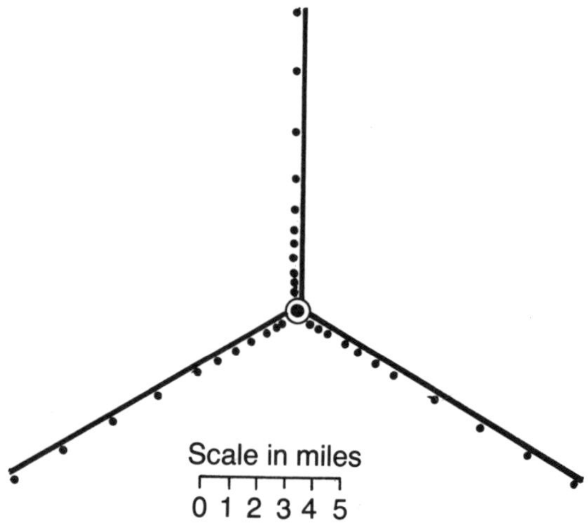

The 27 antennae are spread out on three arms. Two arms are 13 miles long and the other arm is 11.8 miles long.

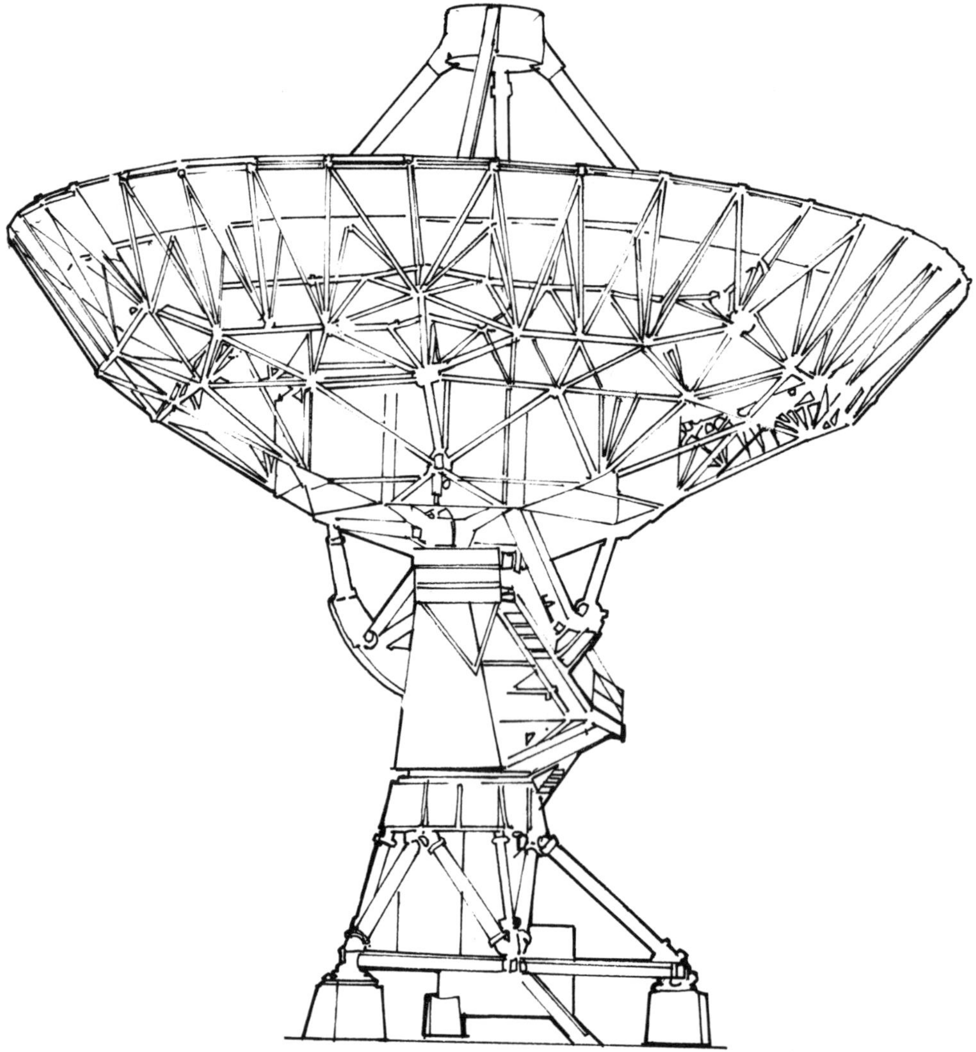

An illustration of one of the 27 giant radio telescopes which make up the National Radio Observatory in Socorro, New Mexico. Each antennae, with its aluminum ear is 82 feet wide and weighs 213 tons.

billion **light years** away. Quasars are the most distant objects "seen" by astronomers. Light has been traveling from some of them for thousands of millions of years. And it is possible that some of them may even be traveling faster than the speed of light. Their size, makeup and how they produce such huge amounts of energy are discoveries still to be made by astronomers.

Other mysteries of the universe to be solved include pulsars and black holes. Pulsars may be stars that rapidly rotate, or "pulsate," and send off the most powerful radio signals in the universe. In black holes, the gravitational pull is so great that no form of light or radiation can escape from them.

New types of optical telescopes are also being used to expand our knowledge of the planets in our solar system, the stars in our galaxy and the other galaxies in the universe. On Mt. Hopkins, in Arizona, a remarkable multiple-mirror telescope has been built. Only the giant 200-inch reflecting telescope on Mt. Palomar and a 236-inch reflecting telescope in the Soviet Union are larger.

The multiple-mirror telescope is made up of six primary mirrors, each 72 inches in diameter. When their images are combined by a computer they are equal in power to a single reflecting telescope with a mirror 176 inches in diameter. By using multiple mirrors, it will be possible to overcome the restrictions caused by the weight and size of a single enormous reflector. The telescope is housed in a 4½-story, 500-ton building that includes offices, labs, workrooms and a lunchroom. The entire structure rotates on steel wheels on a circular track.

The telescope's computer functions much like the human brain and the individual telescopes function like human eyes. Each telescope sees a slightly different image which is transmitted to the computer, where they become one image as they do in the brain. This is accomplished because light from a heavenly body is focused by each

LIGHT YEARS

Unit of measurement for astronomical distance. It is equal to the distance that light covers when traveling in a vacuum for one year, which is approximately 6 trillion or 6,000,000,000,000 miles.

primary mirror, projected to a second mirror and then a third. It is then projected to a beam combiner, and from there the six images are brought into focus. The telescope's computer also makes it possible to merely punch in a list of objects the astronomer wants to observe and the telescope will automatically move from one to the other without much help from the astronomer.

Not only are multiple-mirror instruments much cheaper than a single-mirror reflector, but it is conceivable that it would be possible to combine more than six mirrors and produce gigantic instruments with the light-gathering powers of a 1,000-inch telescope. Such multiple-mirror instruments will primarily be used to study the most distant galaxies. They will also try to produce a more accurate map of our universe.

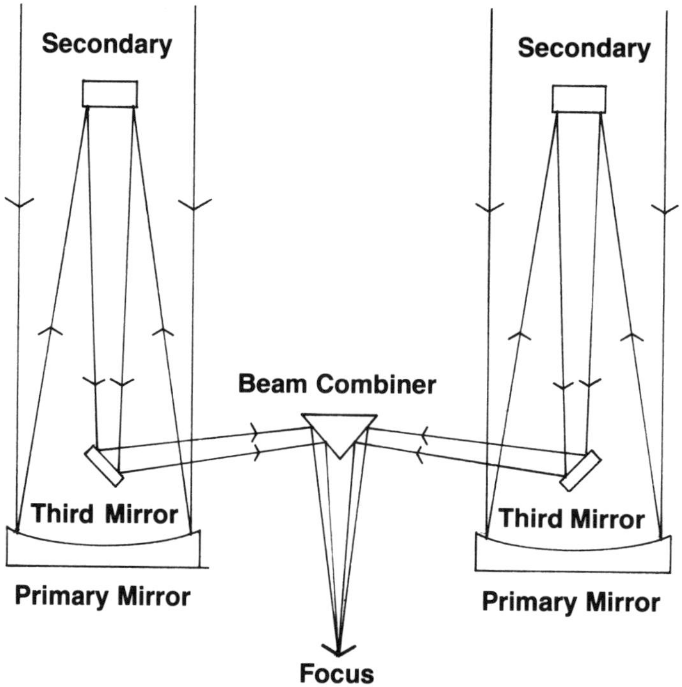

Diagram of how the mirrors in the multiple-mirror telescope function.

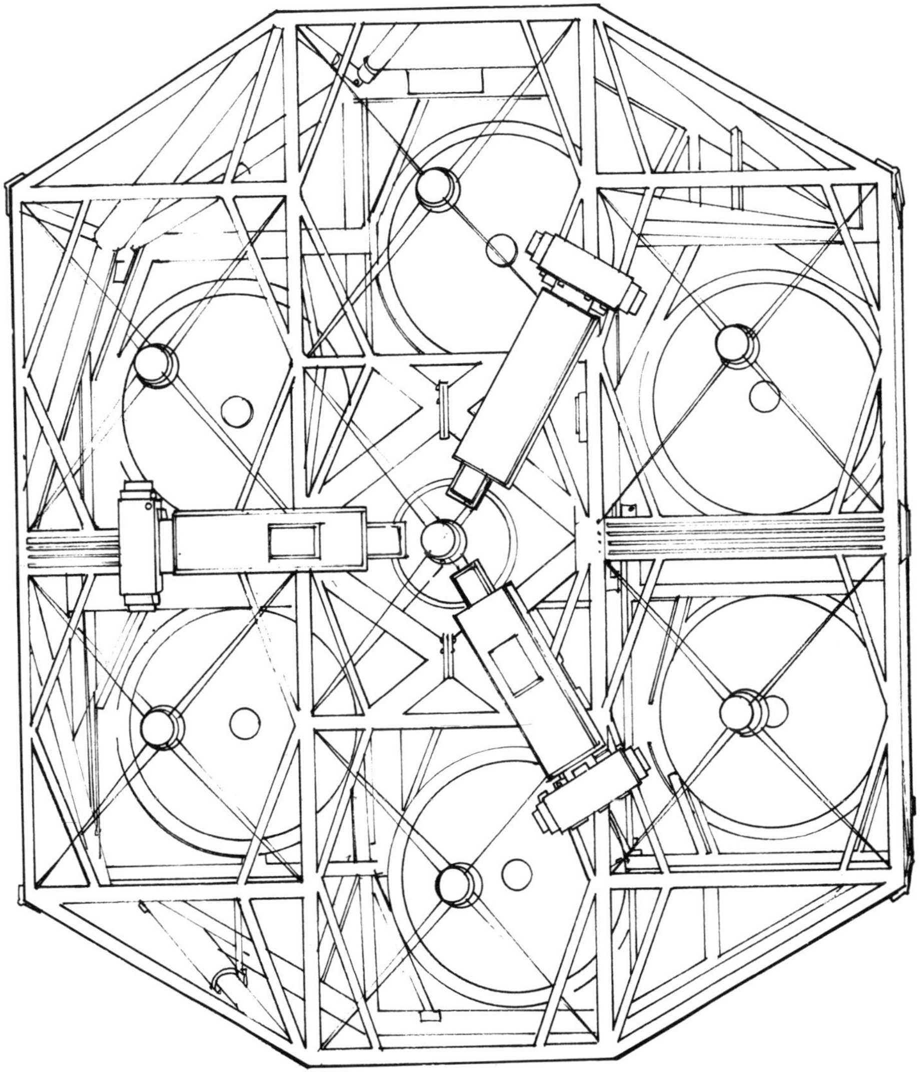

Illustration of the six mirrors which form the "compound eye" of the multiple-mirror telescope on Mt. Hopkins, in Arizona. Each mirror is 6 feet in diameter.

Launch of the Voyager 2 from Cape Canaveral, Florida. *Courtesy National Aeronautics and Space Administration*

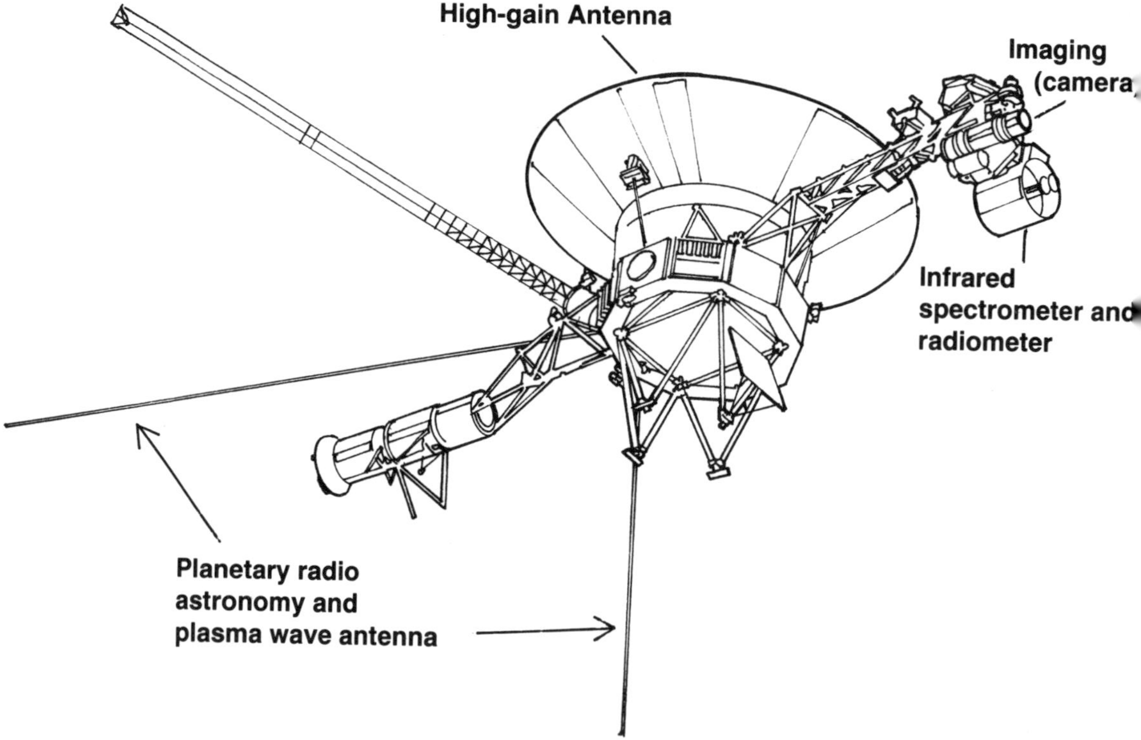

An illustration of a Voyager spacecraft. The spacecraft itself weighs 1,782 pounds and the instruments weigh 232 pounds. The high-gain antenna is 12 feet in diameter.

Equally exciting are the possibilities of carrying telescopes into space away from the interference and disturbances of the Earth's atmosphere. Orbiting optical and X-ray telescopes and astronomical observatories will be able to reveal many of the amazing secrets of our universe.

Photograph of Jupiter and two of its satellites taken on February 13, 1979, from Voyager 1. By that time over 18,000 images had been acquired by the Voyager 1 spacecraft. *Courtesy of the National Aeronautics and Space Administration*

Not only are we using new and exciting astronomical instruments to probe outer space, but we are also sending probes deep into space to photograph the heavenly bodies from close range. In 1977, several Voyager Project spacecrafts were launched from Cape Canaveral, Florida. They are traveling through space to examine Jupiter and its moons close-up, and Saturn and its satellites in a way Galileo probably never could have imagined. These spacecrafts carry such instruments as television cameras, radio telescopes and spectrometers. These instruments will provide us with data about the planets' structures and what they are made of. All of this information is helping us understand our solar system better, and to test the theories about how it began and has evolved.

Since the days when Galileo first observed Jupiter with his simple telescope, we have learned that the universe does not revolve around the Earth. We have come to know our sun as a star in a galaxy of millions of stars. We have discovered that our galaxy is but one of millions of galaxies. We have tried to understand distances that have expanded beyond our imaginations. We have come to realize that the universe is part of a great evolving system that we may never fully understand. Astronomy and astronomical instruments, such as telescopes, will provide us with many new and exciting ways of trying to discover more of the secrets of the universe.